# Springer Proceedings in Earth and Environmental Sciences

**Series Editors**

Natalia S. Bezaeva, The Moscow Area, Russia

Heloisa Helena Gomes Coe, Niterói, Rio de Janeiro, Brazil

Muhammad Farrakh Nawaz, Department of Forestry and Range Management, University of Agriculture, Faisalabad, Pakistan

The series Springer Proceedings in Earth and Environmental Sciences publishes proceedings from scholarly meetings and workshops on all topics related to Environmental and Earth Sciences and related sciences. This series constitutes a comprehensive up-to-date source of reference on a field or subfield of relevance in Earth and Environmental Sciences. In addition to an overall evaluation of the interest, scientific quality, and timeliness of each proposal at the hands of the publisher, individual contributions are all refereed to the high quality standards of leading journals in the field. Thus, this series provides the research community with well-edited, authoritative reports on developments in the most exciting areas of environmental sciences, earth sciences and related fields.

Sebastian Alejandro Avalos Sotomayor ·
Julian M. Ortiz · R. Mohan Srivastava
Editors

# Geostatistics Toronto 2021

## Quantitative Geology and Geostatistics

 Springer

*Editors*

Sebastian Alejandro Avalos Sotomayor
The Robert M. Buchan Department
of Mining
Queen's University
Kingston, ON, Canada

Julian M. Ortiz
The Robert M. Buchan Department
of Mining
Queen's University
Kingston, ON, Canada

R. Mohan Srivastava
RedDot3D Inc.
Toronto, ON, Canada

ISSN 2524-342X                    ISSN 2524-3438  (electronic)
Springer Proceedings in Earth and Environmental Sciences
ISBN 978-3-031-19844-1          ISBN 978-3-031-19845-8  (eBook)
https://doi.org/10.1007/978-3-031-19845-8

This Springer imprint is published by the registered company Springer Nature Switzerland AG
The registered company address is: Gewerbestrasse 11, 6330 Cham, Switzerland

# Preface

The world was a much simpler place on that sunny day in Valencia in September 2016 when the geostat community chose to meet in Toronto for the next meeting in the Geostatistics Congress series. We lined up wonderful meeting space at Victoria College on the University of Toronto grounds, had an engaging group of invited keynote speakers and a beautiful location for the conference dinner on the lakefront… what a great meeting it would have been, but for a new bundle of genetic material that is surrounded, like a crown, with spike proteins.

When Oy Leuangthong and I met in February 2020 to make the final selection of papers for oral and poster presentations, we knew what the "novel coronavirus" was but we had no idea that barely a month later, the world would shut down and that we would all learn to adapt to working from our kitchen tables, become familiar with Zoom meetings and with terms like "social distancing", "anti-vaxx", "flattening the curve" and "asymptomatic transmission".

As the lead organizer for the 11th Geostatistics Congress, I faced a tough decision when the global pandemic was declared in March 2020: run the conference as planned, postpone it or cancel it. As the weight of a global pandemic was being felt everywhere, some counseled me to stay the course: the wave of infections would clear by May, I was told, and the world would be suitably back to normal by July… yeah, right. Even in the week of the declaration of the pandemic, that rosy optimism seemed unwarranted, and postponing the conference to 2021 seemed like the only viable option. But by the late Spring of 2021, even with vaccines available, we were starting to learn about Delta variants. Travel was still severely restricted and, in Toronto, large indoor meetings were not yet permitted, leaving no good choices for the meeting we wanted to have.

Postponing yet again, to 2022, wasn't a happy thought. With 16 months having passed under pandemic restrictions, it was not at all clear that the passage of another year would make an in-person meeting possible. If we missed 2022, and rolled forward to 2023, it would be seven years since the previous Geostatistics Congress, and my concern was that the relevance of the series would come into question. This has been a remarkable series, not only for its longevity… coming up to 50 years… and the quality and breadth of its contributions to geostatistics, but also for its ability to create a sense of a community. For many of us, meeting with colleagues once every four years was an important piece of social glue. For young professionals these meetings have been an opportunity to network and to present their work to future employers and colleagues. For those of us who remember all the way back to Frascati and Lake Tahoe, these were a chance to reconnect, over drinks and meals, with the many bright and stimulating people who choose to call themselves geostatisticians. And for everyone, these meetings have been a chance to share ideas, to learn new tricks and to share our thoughts on the successes and failures we've encountered on the bridge from theory to practice.

As geostatistics has proven its value in many fields, the breadth of applications has widened, leaving us in the situation of often not understanding what other geostatisticians are doing. Specific knowledge necessary to understand problem-solving in one area of application is often a complete mystery to people who focus on a different area. Mining geostatisticians go glassy-eyed when petroleum geostatisticians talk about relative permeability curves or hysteresis. The petroleum folks have a hard time making sense of grade-capping or open-pit production optimization. And the others who ply their trade as geostatisticians in environmental studies, or epidemiology, or agriculture or animal census studies… they just shake their head and wonder if they're in the right room. This increasing disparity in domain-specific knowledge continues and leaves many geostatisticians feeling that maybe the next Geostatistics Congress is not the best use of their conference time. It was against this backdrop that I worried about allowing seven years to go by since we last met in Valencia. If we did fine for seven years without a Geostatistics Congress, maybe the series had reached its logical end and we should all scatter into specialized geostatistics sessions in other conferences. For me, having begun my professional career around the time this series began, that was a sad thought. Perhaps it is inevitable, but I didn't want it to be on my watch.

If I wasn't going to delay to 2022, and risk perhaps delaying again to 2023, the only alternative was to hold the conference online as a virtual conference, an option that had its own clear drawbacks. During the pandemic, we all learned to dislike Zoom-style conferences with their techno-glitches, their lack of direct contact, the loss of opportunities for socialization and the wrangling of appropriate meeting times for people in 24 time zones. But even though an online meeting was not what any of us looked forward to, I believed that an online conference could work well for several reasons:

- Presenters could be told to prepare their presentations beforehand, as videos, which gave everyone a chance to rehearse and polish their presentations.
- Pre-recorded presentations also made time-management a lot easier … we just sped up the clip speed for those few people who did not make their 20-minute mark.
- Anyone interested in a particular talk could review the presentations ahead of time, freeing up the time in the week of the live online conference for extended question-and-answer discussion sessions.
- People from around the world could participate, including those who do not have the means to travel to a distant conference venue for a week.
- We wouldn't have to split presentations into oral and poster sessions.
- Everyone could participate in the same sessions; there would be no need for parallel break-out sessions.

Although many people who had heir abstracts accepted decided to opt out, about 60 decided to participate as presenters in the online version. And they did very well, not only with their presentations but also with stimulating discussions during the live Zoom meetings, and with their answers to questions posted to a bulletin board. The live discussion sessions were held at times that might work for many of the participants, from morning on the west coast of North America to late afternoon in Europe. But there were also several presenters and participants who were up in the wee hours of the morning in Australia and Asia.

Command central was my kitchen table, and my aide-de-camp in the ground war of registrations, video editing, websites, bulletin boards and Zoom troubleshooting was my teenage son, Ravi. We had no idea what to expect when we went live on the morning of Monday, July 12, 2021. At that point, we had just over 100 registrations, but they were still trickling in... so Ravi and I tag-teamed on making sure that new registrants had access to the pre-recorded talks, the abstracts and the bulletin board. And we also had to monitor live sessions, sort out the mid-day invited speakers and make sure that a couple of online workshops ran correctly.

Between the two of us, Ravi and I managed to launch on Monday morning but we hadn't counted on a wave of new registrations that soon turned into a tidal wave. At first, I was surprised because I expected that anyone who would want to participate as a spectator, without being involved in any presentation, would have registered before the start of the conference. But as the registrations started to climb quickly on Monday, it was apparent that we were getting word-of-mouth recommendations from people spreading the word to others in their same organization. Multiple registrations came in from the same companies, or the same research centers. My best guess on what happened was that a lot of people took a wait-and-see attitude, waiting to see if this online meeting was as painful as most others that we've suffered through. As the morning sessions unfolded, and participants realized that the discussions were well-organized, interesting and informative, they passed on the word that this was worth spending time on and suggested to their colleagues that they join in. Within a couple of days, the number of registrations climbed to over 200, which is about the number of people who have typically registered for past congresses in this series.

We still missed out on the drinks and the meals together, but we did manage a bit of social activity: an online "5k race" done using Google Maps, and an art show. And the discussions were lively, with some long and insightful exchanges about the philosophy of modeling, and that tug-of-war between computer power and human reason. My sense is that the online format actually encouraged participation in the Q&A sessions, not only because some people are more comfortable when they're not standing up in front of the world's elite geostatisticians but also because many of the people who participated were those who do not have the resources to travel to attend major international conferences. There are graduate students, young practitioners, teachers and researchers around the world who are using geostatistics but who have never had the chance to talk with people who write the papers they've read. They thirst for knowledge, both practical and theoretical, and have a keen curiosity about why some things work well and some things don't. The online format gave many people a chance to exercise their curiosity and, for the first time in their professional careers, to speak with geostatisticians from around the world. Having worried that this series might be reaching the end of its relevance, I was pleased to see that it hadn't. For many people who normally would not be able to attend, the 11th International Geostatistics Congress filled a need.

These proceedings include all of the presentations that were made. The presenters had several choices for what to include in this volume. Many chose to stick with their original abstract, which gives a quick overview of the work they presented. Some chose to submit an extended abstract, a few pages that allow them to expand on their original one-page abstract by expanding on the discussion and adding some figures and tables. Some chose to submit a full paper that was peer-reviewed.

The journal Mathematical Geosciences has, in the past, devoted one of its Special Issues to selected papers from previous Geostatistics Congresses. They are doing so again, with six of the papers from Geostats 2021 having been adapted to their journal format, and peer-reviewed again to meet their requirements for scientific quality. For these papers, the proceedings contain an abstract or extended abstract; the full papers, updated to include additional content, will appear in a Mathematical Geosciences Special Issue later this year.

The videos of the presentations are all still accessible on the geostats2021.com website. The registration requirement has been removed, and anyone who wants to see the full original presentations can access the videos.

The range of topics covered in these proceedings is a testament to the value of geostatistics. A quick skim of the table of contents will confirm that the theoretical tools that first found application in the mining industry have evolved and been adapted to the needs of many other applications. And, as happens with flexible tools that have enjoyed success in practice, geostatistics now overlaps with other disciplines. There are many presentations here, for example, that explore the use of artificial intelligence and machine learning in geostatistical studies.

One of the traditions that has survived through eleven Geostatistics Congresses is that the structure is remarkably informal. There is no professional body that oversees this series. The people who participate in each meeting choose the next venue and turn over the organization to a volunteer committee. At the end of the Toronto meeting, the participants voted for the next in the series to be held in the Azores in 2024. I know that we will all look forward to getting back to an in-person meeting, but I also hope that we again find a way to be accessible to those who cannot afford to attend. I know from that week-long blur in mid-July last year that many people find value in geostatistics as a body of knowledge in its own right, and not merely as a useful toolkit in specific areas of practical application.

I am grateful to the many who made the 11th International Geostatistics Congress a success, through their participation as presenters, as session chairs and as active participants in the discussion sessions. Despite the many difficulties imposed by a global pandemic, we succeeded in coming together once again to exchange ideas, and to enjoy the intelligence, wisdom and laughter of this community. Many thanks to all.

Toronto, Canada                                                      R. Mohan Srivastava
                                                                          Chairman, 11th International
                                                                          Geostatistics Congress

# Acknowledgements

The organizers of the 11th International Geostatistics Congress wish to thank the dozens of corresponding authors who put a lot of work into their presentations and who went the extra mile when they were told that they had to pre-record and upload videos a week before the conference AND keep them to within 20 minutes. The quality of their work, their cooperation, their patience and their good humor are all very much appreciated.

The centerpiece of each day of the online conference was a talk from an invited speaker who was asked to reflect on things that worked well and things that didn't ... a rather open-ended assignment that they each handled with flair and their own personal spin. Geostats 2021 owes much to Peter Atkinson, Colin Daly, Jaime Gómez-Hernández and Melanie Stefan, whose invited keynote talks were essential to creating a sense that we were at a "real" conference.

The heavy lifting on the preparation of the proceedings has been ably handled by Sebastian Avalos, with the assistance of a technical committee who reviewed papers:

Avalos, Sebastian, Queen's University, Canada
Boisvert, Jeff, University of Alberta, Canada
da Silva, Camilla, University of Alberta, Canada
Dawar, Kshitij, The Pennsylvania State University, USA
Emery, Xavier, University of Chile, Chile
Fouedjio, Francky, Rio Tinto, Australia
Harding, Ben, University of Alberta, Canada
Lantuejoul, Christian, School of Mines ParisTech, France
Liu, Wendi, The University of Texas at Austin, USA
Madani, Nasser, Nazarbayev University, Kazakhstan
Manzocchi, Tom, University College Dublin, Ireland
Mery, Nadia, University of Chile, Chile
Morales, Daniel, McGill University, Canada
Muller, Ute, Edith Cowan University, Australia
Oliveira, Thyago, AngloGold Ashanti, Brazil
Ortiz, Julian, Queen's University, Canada

Riquelme, Alvaro, Queen's University, Canada
Silversides, Katherine, The University of Sydney, Australia
Tolosana-Delgado, Raimon, HZDR, Germany
Yao, Lingqing, McGill University, Canada.

Finally, behind the scenes since 2016 was a committee who organized two complete versions of the 11th International Geostatistics Congress, Geostats 2020 and Geostats 2021. Only one of these saw the light of day, but the time and energy of Ravi Butler, Oy Leuangthong, Dolly Reisman and Mohan Srivastava on both versions are greatly appreciated.

Editorial Committee
11th International Geostatistics Congress

# Remembering Dr. Harry M. Parker (1946–2019)

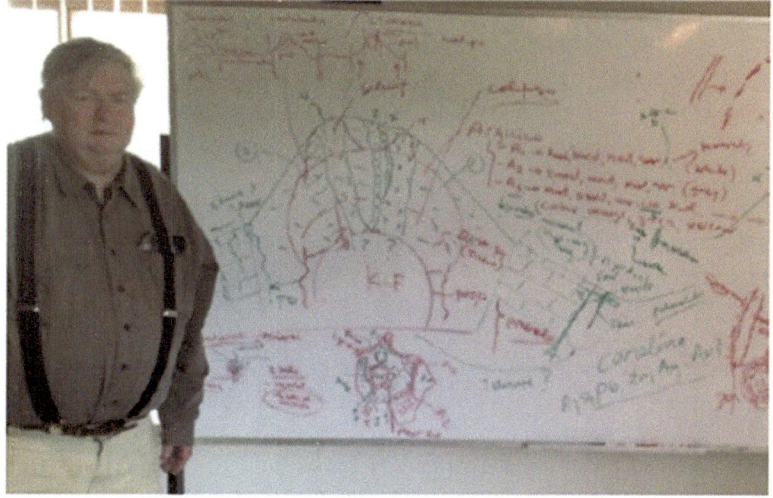

Shortly before the time when the 11th International Geostatistics Congress was originally scheduled, the world of geostatistics lost one of its pillars: Harry Parker. The plans for Geostats 2020 included a session focused on Harry's many contributions, much as we did in 2016 when we noted the passing of Danie Krige. It's a gentle and thoughtful tradition: remembering the giants on whose shoulders we stand.

The success of geostatistics owes much to Harry, who actively advocated its use in his mining consulting work and who was instrumental in the careers of many geostatisticians. When André Journel moved to North America in the mid-1970s to take up a visiting professor position at Stanford University, Harry secured André a half-time consultancy with Fluor Mining and Metals where Harry ran the Geology and Geostatistics group. This gave André reason to stay until Stanford offered him a position as a full professor. In the late 1970s, Harry interviewed a young Mo Srivastava in his senior year at MIT and convinced him that geostatistics, which Mo had never heard of, included the possibility of kriging on the coral beaches of Cebu in the Philippines.

He promoted the use of conditional simulation years before it became widely accepted, and is a big part of the reason it did become widely accepted. Harry was using geostatistical methods for the estimation of recoverable reserves when much of the mining industry remained content with polygons and inverse distance interpolation. Even though he was, in a sense, decades ahead of his time, Harry cultivated the image of the old-school geologist, pretending to be skeptical of equations and statistics even though he had mastered geostatistics during his Ph.D. work. His doctoral thesis was the second Ph.D. awarded in geostatistics in the United States.

Even though he liked to play the role of the gruff geologist, Harry was completely at home in the world of geostatistics. He was one of the last people to have attended every Geostatistics Congress since Frascati. Harry authored and co-authored more than 40 technical papers and presentations over the course of his career. His papers touched on many of the most important aspects of resource estimation, including geological controls, outliers, analysis of spatial continuity, recoverable reserves, conditional simulation, production reconciliation, risk analysis and quantification of uncertainty. His 1978 paper on the volume-variance relationship remains one of the clearest explanations for why a mineral deposit's grade and tonnage above a cutoff grade have to depend on the level of selectivity.

In addition to his many contributions to the theory and practice of geostatistics, Harry was known for his generous spirit. He was glad to spend time with juniors and peers, helping others better understand advanced methods and passing on his philosophy that incorporating geological controls was the key to building reliable numerical models of the subsurface. He was meticulous in checking data and in checking the correctness of software implementations, skills that he passed on to a generation of geologists and resource estimation specialists who were mentored by him.

He was awarded the AusIMM's Institute Medal in 2019, the SME's President's Citation in 2017, the SME's Award for Competence and Ethics in 2012, and the Southwest Mining Foundation's American Mining Hall of Fame Medal of Merit in 2007. He was made an Honorary Life Member of the Geostatistical Association of Australia, an Honorary Fellow of the Professional Society of the Independent Subsoil Experts of Kazakhstan, and an Honorary Representative of CRIRSCO, the international organization that oversees and is the harmonization of international reporting standards for mineral resources and reserves. He also was conferred the Mongolian Best Geologist Award in 2015, and given APCOM Recognition Awards in 2015 and 2017. He was posthumously awarded the Harry M Parker Excellence Award and AIME/SME Saunders Award.

# Contents

# Theory

# A Geostatistical Heterogeneity Metric for Spatial Feature Engineering

Wendi Liu, Léan E. Garland, Jesus Ochoa, and Michael J. Pyrcz

**Abstract** Heterogeneity is a vital spatial feature for subsurface resource recovery predictions, such as mining grade tonnage functions, hydrocarbon recovery factor, and water aquifer draw-down predictions. Feature engineering presents the opportunity to integrate heterogeneity information, but traditional heterogeneity engineered features like Dykstra-Parsons and Lorenz coefficients ignore the spatial context; therefore, are not sufficient to quantify the heterogeneity over multiple scales of spatial intervals to inform predictive machine learning models. We propose a novel use of dispersion variance as a spatial-engineered feature that accounts for heterogeneity within the spatial context, including spatial continuity and sample data and model volume support size to improve predictive machine-learning-based models, e.g., for pre-drill prediction and uncertainty quantification. Dispersion variance is a generalized form of variance that accounts for volume support size and can be calculated from the semivariogram-based spatial continuity model. We demonstrate dispersion variance as a useful predictor feature for the case of hydrocarbon recovery prediction, with the ability to quantify the spatial variation over the support size of the production well drainage radius, given the spatial continuity from the variogram and trajectory of the well. We include a synthetic example based on geostatistical models and flow simulation to show the sensitivity of dispersion variance to production. Then we demonstrate the dispersion variance as an informative predictor feature for production forecasting with a field case study in the Duvernay formation.

W. Liu (✉) · M. J. Pyrcz
Hildebrand Department of Petroleum and Geosystem Engineering, Cockrell School of
Engineering, University of Texas at Austin, Austin, USA
e-mail: wendi_liu@utexas.edu

L. E. Garland
Equinor, Stavanger, Norway

J. Ochoa
Equinor, Houston, USA

M. J. Pyrcz
Department of Geological Sciences, Jackson School of Geosciences, University of Texas at
Austin, Austin, USA

S. A. Avalos Sotomayor et al. (eds.), *Geostatistics Toronto 2021*, Springer Proceedings
in Earth and Environmental Sciences, https://doi.org/10.1007/978-3-031-19845-8_1

**Keywords** Feature engineering · Machine learning · Heterogeneity ·
Geostatistics · Unconventional reservoir

# 1  Introduction

Feature engineering is the formulation and compilation of a set of informative
predictor features from raw data features to improve the performance of the predictive machine learning and other statistical models [1, 2], which includes selecting
the most relevant available predictor features, removal of redundant features and
constructing new predictor features. For example, the engineered predictor features
based on local gradient filters are effective at highlighting data trends, and edge detection filters are effective at finding critical boundaries or transitions in data; therefore,
it is common to use these engineered features in applications such as computer vision.
Raw features may be combined to construct new features, for a specific subsurface
case consider rock quality index as a combination of rock permeability and porosity
[3]. The rapid development of deep learning shifts the burden of feature engineering
to its underlying learning system [4], such as convolutional neural networks that
learn local image transformations represented as weighted filters known as kernels.

Spatial feature engineering is critical for subsurface machine learning applications
because deep learning captures spatial features learned from dense data representations like images or time series but the spatial hard data available for understanding the
subsurface volume of interest (e.g., a hydrocarbon/water reservoir or mining deposit)
are sparse, such as core measurements or well logs. Also, the spatial context of the
subsurface data, such as sampling manner, spatial continuity and multiple scales,
known as support size, of data and models, obscure the relationships between subsurface predictor features (e.g., rock type, porosity, and permeability) and response
features (e.g., flow rates and mineral grade) for data-driven prediction model performance, and must be dealt with prior to feature engineering. Efforts have been made to
address the spatial sampling issues for spatial machine learning model construction,
including spatial sampling bias [5], spatial anomalies [6], spatial training and testing
data splitting [7] and spatial statistical significance [8]. It is essential to capture other
aspects of the spatial context, i.e., the spatial feature heterogeneity and the scale,
volume support size of the data, in model predictions. Heterogeneity is the variation
in subsurface features as a function of spatial location and is a vital factor to predict
subsurface resource recovery [9]. Also, subsurface datasets and models span a large
range of scales from well and drill hole cores, core plugs, well logs, remote sensing
and production or recovery data sources. A practical heterogeneity metric can be
applied as a spatial-engineered feature for inputs into predictive models to improve
the integration of the spatial context to potentially improves the model performance.

Common non-spatial heterogeneity metrics include Dykstra-Parsons and Lorenz
coefficients [10–12], which are relatively easy to estimate without much computational power. However, these metrics may be calculated from the permeability and
porosity data table and ignore the spatial context like location, spatial continuity

and support size. There are proposed heterogeneity metrics that attempt to integrate spatial correlation in, including Polasek and Hutchinson's heterogeneity factor [13] or Alpay's sand index [14], but these metrics require extrapolation of the local hetero-geneity measurements based on fence diagram or sand isopach map, which is too smooth. There are various research efforts focusing on the comprehensive analysis of heterogeneity characterization that is informative for the specific support size or type of reservoir [15–17] but lack the flexibility and computational efficiency to be generalized as a spatial feature for predictive machine learning models.

Dispersion variance is a generalized form of variance that accounts for the volume support size of data samples and models [18, 19]. It is relatively fast to calculate based on well or drill hole measurements while integrating spatial continuity. Dispersion variance is denoted as $D^2(a, b)$, representing the variance of the feature of interest measured at volume support size a in the larger volume b. Dispersion variance honors the additivity of variance relation, which is the foundation of the analysis of variance (ANOVA) in statistics. For example, the scale change from core scale to geological modeling scale, consider as the volume support of the spatial sample data (denoted as ·), as the support size of the geological modeling cells (denoted as $v$), while the volume of interest is denoted as $V$. Then the dispersion variance can be decomposed into the following based on the additivity rule, known as Krige's relation:

$$D^2(\cdot, V) = D^2(\cdot, v) + D^2(v, V) \tag{1}$$

the total variance of the samples in the volume of interest is equal to the sum of the variance of the samples in the geological model cells and the variance of the geological model cells in the volume of interest.

The dispersion variance of different volume support sizes can be estimated from the volume integrated semivariogram $\gamma(\mathbf{h})$, denoted as $\overline{\gamma}(\mathbf{h})$ and stated as 'gamma bar', where one extremity of the vector $\mathbf{h}$ describes the domain $v(\mathbf{u})$ and the other extremity independently defines domain $V(\mathbf{u}^*)$. Semivariogram $\gamma(\mathbf{h})$, under the assumption of stationary of the mean and variogram, is defined as:

$$2\gamma(\mathbf{h}) = \mathbb{E}\{[Z(\mathbf{u} + \mathbf{h}) - Z(\mathbf{u})]^2\}, \forall \mathbf{u}, \mathbf{u} + \mathbf{h} \in AOI \tag{2}$$

where $2\gamma(\mathbf{h})$ is the variogram, $\mathbf{u}$ is the coordinate location vector, $\mathbf{h}$ is the lag distance vector separating all pairs of data, $Z(\mathbf{u})$ and $Z(\mathbf{u} + \mathbf{h})$ of Z variable and AOI is the area of interest. Semivariogram is half of the variogram. Here we adopt the common short-hand of variogram to represent the semivariogram. In order to calculate dispersion variance from volume integrated variogram, a permissible, positive definite, para-metric nested variogram model informed by the experimental variogram is needed to provide a continuous function that is valid for all distances and directions, as:

$$\gamma(\mathbf{h}) = \sum_{i=1}^{nst} C_i \Gamma_i(\mathbf{h}) \tag{3}$$

where $nst$ is the number of nested variogram functions, $C_i$ is the variance contribution of each nested structure and $\Gamma_i(\mathbf{h})$ is the variogram function for each variance contribution acting over all distances and directions modeled in the major and minor continuity directions and interpolated for all other directions with a geometric anisotropy model. Then the volume integrated variogram, gamma bar $\overline{\gamma}(\mathbf{h})$, is calculated by:

$$\overline{\gamma}\big(v(\mathbf{u}),\ V(\mathbf{u}^*)\big) = \frac{1}{v \cdot V} \int_v \int_V \gamma\big(y - y^*\big) dy dy^* \tag{4}$$

where $v$ and $V$ are volumes of $v(\mathbf{u})$ and $V(\mathbf{u}^*)$. Then the dispersion variance can be estimated from $\overline{\gamma}$ as:

$$D^2(v,\ V) = \overline{\gamma}(V,\ V) - \overline{\gamma}(v,\ v) \tag{5}$$

This volume-variance relation has been proven powerful in reconciling feature variance across multiple scales and volume support sizes while accounting for spatial continuity [20, 21]. Therefore, it is a reasonable hypothesis that dispersion variance could be a good spatial feature for predictive models by integrating heterogeneity for different volume support sizes.

We propose a novel utilization of dispersion variance as a heterogeneity metric for spatial feature engineering for subsurface predictive machine learning. The proposed heterogeneity metric is not only practical to calculate as common static heterogeneity metrics, but also integrates the spatial continuity and the volume support size of the predictor features, which are critical aspects of the spatial context [22]. We demonstrate that our proposed engineered feature is sensitive to variations over a variety of spatial settings while remaining easy to compute and that application as a new predictor feature improves machine learning prediction performance for the case of hydrocarbon recovery from a heterogeneous reservoir. The sensitivity of the proposed engineered feature (i.e., dispersion variance within well drainage radius) is shown by changing with possible variables, such as well length, well drainage radius, and well trajectory. Then we model variogram, map the volume of the data from the varying well trajectories and investigate the hydrocarbon recovery information informed by dispersion variance.

In the next section, we explain the methodology for practical calculation of our proposed dispersion variance-based heterogeneity spatial-engineered feature. In the results section, firstly, we show the sensitivity analysis results from the possible factors that affect dispersion variance calculation from the perspective of geological stratigraphy and well parameters. Secondly, we investigate the relation between dispersion variance and hydrocarbon production based on black-oil simulation while controlling other simulation parameters to be constant. Lastly, we conduct a case study with data from the Kaybob field, Duvernay Formation, where we use dispersion variance as a spatial feature with other completion and petrophysics features for machine learning models to demonstrate that dispersion variance is an informative spatial-engineered feature for hydrocarbon recovery predictive model.

## 2 Methodology

Our proposed method assumes stationary mean, $\mu_z$, variance, $\sigma_z^2$, and variogram $\gamma_z(\mathbf{h})$ of spatial features over the area of interest (AOI). Given this assumption of stationarity, for the calculation of the variogram, we omit the dependence on location and only consider the dependence on the lag vector, $\mathbf{h}$.

$$\gamma(\mathbf{h}; \mathbf{u}) = \gamma(\mathbf{h}) \qquad (6)$$

In the presence of non-stationarity, we can model a local trend model and work with a stationary residual or segment the area of interest into multiple stationary regions with domain expertise.

The steps to calculate the dispersion variance-based heterogeneity spatial-engineered feature are:

1. Calculate the representative feature distribution variance at data support. Declustering methods can be utilized to achieve the goal of statistical representativity by assigning each datum a weight [23, 24]. The representative feature distribution variance $\sigma_z^2$ can be approximated from sample variance $s_z^2$ by:

$$s_z^2 = \sum_{i=1}^{n} w_i (z_i - \bar{z})^2 \qquad (7)$$

where weight $w_i, i = 1, 2, \ldots, n$ are between 0 and 1 and add up to 1, $z_i$ is the sample datum, $\bar{z}$ is the representative sample mean calculated from:

$$\bar{z} = \sum_{i=1}^{n} w_i z_i \qquad (8)$$

2. Calculate and model the variogram integrating all available spatial data, analog information and trend model.
3. Establish the volume support size of the sample data, or imputed data, $v$, volume. For the case of predicting hydrocarbon or water recovery from a well, the volume support size is based on the well drainage radius and the well length. The variance of data within the well volume is denoted as $D^2(\cdot, v)$, where $\cdot$ is the volume support of the spatial sample data, $v$ is the volume within well drainage radius, i.e., $D^2(data, well)$. According to Eq. 1:

$$D^2(data, reservoir) = D^2(data, well) + D^2(well, reservoir) \qquad (9)$$

The dispersion variance can be calculated from a volume-integrated variogram, $\overline{\gamma}(\mathbf{h})$ according to Eq. 4. The dispersion variance within the volume of drainage radius of a horizontal well is calculated with $\overline{\gamma}(\mathbf{h})$ according to Eq. 5 as follows:

$$D^2(data, well) = \overline{\gamma}(well, well) - \overline{\gamma}(data, data) \qquad (10)$$

4. Apply numerical integration to calculate the dispersion variance of the feature over the volume support $v$ of the sample data and apply it as a new engineered feature. The numerical approximation for $\overline{\gamma}(\mathbf{h})$ can be estimated as:

$$\overline{\gamma}\big(v(\mathbf{u}), V\big(\mathbf{u}^*\big)\big) \approx \frac{1}{m \cdot m^*} \sum\nolimits_{i=1}^{m} \sum\nolimits_{j=1}^{m^*} \gamma\big(\mathbf{u}_i - \mathbf{u}_j^*\big) \tag{11}$$

where the $m$ points $\mathbf{u}_i$, $i = 1, 2, \ldots, m$ discretize the volume $v(\mathbf{u})$ and the $m^*$ points $\mathbf{u}_j$, $j = 1, 2, \ldots, m^*$, discretize the volume $V(\mathbf{u}^*)$. For calculating $\overline{\gamma}(well, well)$ with a given volume support of the well, the volume support size $v(\mathbf{u}) = V(\mathbf{u}^*)$, which is the volume within the well drainage radius. In $\overline{\gamma}$ calculation, we assume the variogram of the feature average linearly within area of interest. If the original feature is standardized, and under the standardized scale with the stationary assumption, $\overline{\gamma}(data, data) = 0$, and $\overline{\gamma}(reservoir, reservoir) = \sigma_z^2 = 1$. So, the standardized dispersion variance is between 0 and 1. We will use standardized dispersion variance throughout our demonstrations.

From this workflow, we calculate the proposed dispersion variance of the volume support size and then apply it as a spatial-engineered feature with improved integration of the spatial context.

## 3  Results and Discussion

Based on the above workflow steps of calculating the spatial engineering features, dispersion variance at a given support volume size, the possible factors that affect dispersion variance are well length, well drainage radius, well trajectory (i.e., dip and azimuth deviating from the major direction of the spatial continuity model). We demonstrate the impact of each factor on dispersion variance within well drainage radius $D^2(\cdot, v)$ through a sensitivity analysis first. Then we demonstrate $D^2(\cdot, v)$ as a spatial-engineered feature for hydrocarbon production prediction models. Note, the features are standardized for standardized dispersion variance that is bound between 0 to 1.

To demonstrate the proposed heterogeneity metric as an engineered feature, we build a 3-dimension 590 m × 590 m × 90 m heterogeneous hydrocarbon reservoir model as a truth model based on sequential Gaussian simulation for primary variable porosity and with collocated cokriging for secondary variable logarithm permeability [25, 26]. Both porosity and permeability in the logarithm scale have the same variogram model and a 0.7 correlation coefficient. The variogram model parameters are in Table 1. The truth reservoir model is shown in Fig. 1.

We demonstrate the sensitivity of the spatial-engineered feature with respect to well length, drainage radius and well trajectories. Figure 2 shows a schematic indicating the spatial-engineered feature volume support size and variogram model

**Table 1** Variogram model parameters to generate heterogeneous reservoir model

| Nugget effect = 0 | Number of nested structures = 2 | | Azimuth = 45° | | Dip = 0° |
|---|---|---|---|---|---|
| Structure | Type | Variance contribution of each nested structure | Major direction range (m) | Minor direction range (m) | Vertical direction range (m) |
| First | Spherical | 0.5 | 400 | 200 | 10 |
| Second | Spherical | 0.5 | 800 | 500 | 40 |

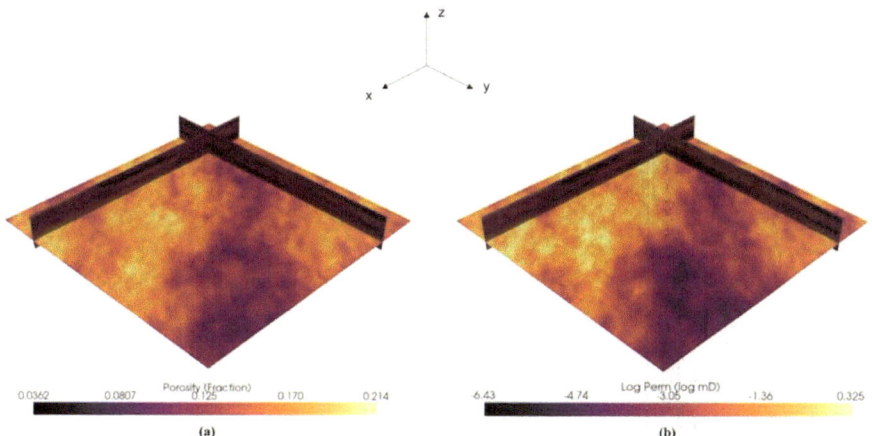

(a)  (b)

**Fig. 1** Truth reservoir model sections including (**a**) porosity and (**b**) permeability in logarithm scale

parameters for the corresponding well trajectories. The volume support size is quantified by well drainage radius and well length. For well trajectories, we are interested the dip angle and azimuth deviating from major directions of the geological spatial continuity specified in the variogram model. Figure 3 shows an illustration of the proposed spatial-engineered feature, dispersion variance over the well drainage volume, as a function of the deviation of the well trajectory from the major spatial continuity azimuth and dip direction with fixed well length and well drainage radius. The well trajectory that is aligned with the major direction and the stratigraphic layer, i.e., when $\Delta$ dip $= 0$ and $\Delta$ azimuth $= 0$, has the minimum dispersion variance within drainage radius $D^2(\cdot, v)$, as larger variogram range decreases $D^2(\cdot, v)$ and increases $D^2(v, V)$, dispersion variance between wells. When well trajectory aligns with the major direction and $\Delta$ dip $= 0$, the variogram range within well drainage radius is maximized. Figure 4 demonstrates the sensitivity of standardized dispersion variance within well drainage radius with respect to well length, well drainage radius and well trajectories. Table 2 shows the factors values for the base case and their range for test cases applied to calculate the sensitivity shown in Fig. 4. The base case

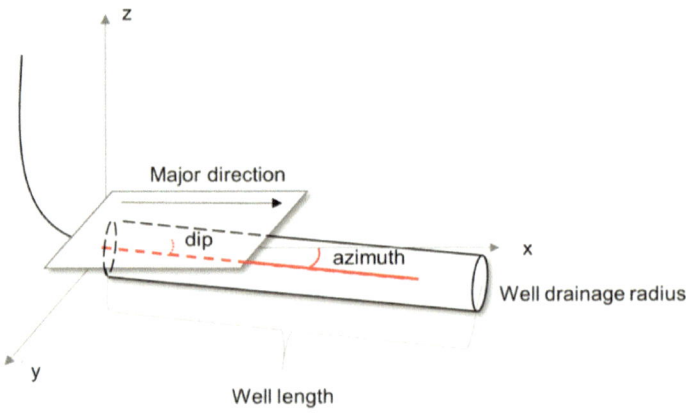

**Fig. 2** Schematic illustration indicating the spatial-engineered feature volume support size and variogram parameters for the corresponding well trajectories

parameters and test case ranges are picked based on the major and minor direction range of the variogram model. The base case has a standardized dispersion variance value $D^2(\cdot, v) = 0.696$. Well length has the largest impact on dispersion variance value in comparison with all other factors.

Next, we demonstrate the ability of the spatial-engineered feature to predict subsurface hydrocarbon production behavior with flow simulation. We construct a black-oil, finite difference, implicit pressure explicit saturation (IMPES) numerical simulation model with the truth reservoir model in Fig. 1. The production relies on pressure depletion only, to simplify the production forecast simulation so that we can focus on the impact of heterogeneity.

Since well length and well drainage radius have an obvious impact on production, to investigate the impact of the dispersion variance as the proposed spatial feature on hydrocarbon production, we assume the constants for well length and well drainage radius as the base case values; therefore, dispersion variance only changes with respect to well trajectories. We iterate over realizations of varying dip and azimuth of well trajectory and calculate the corresponding dispersion variance within well drainage radius while controlling other simulation parameters to be the same for each realization. Then for each well trajectory in the realization, there is a corresponding cumulative production curve over time. We group the cumulative production curves using the base case $D^2(\cdot, v)$ value as cut-off for high and low $D^2(\cdot, v)$. Figure 5 shows the 95% confidence interval of the production curve conditional to the dispersion variance within the drainage radius of the well, which indicates the significantly different cumulative oil production with low and high $D^2(\cdot, v)$. Therefore, the dispersion variance is informative as a spatial-engineered feature to be utilized in the data-driven predictive model for production.

Additionally, a sensitivity analysis is conducted to investigate the flow performance over various permeability magnitudes and permeability heterogeneity quantified by the Dykstra-Parsons coefficient. We use the relative cumulative production

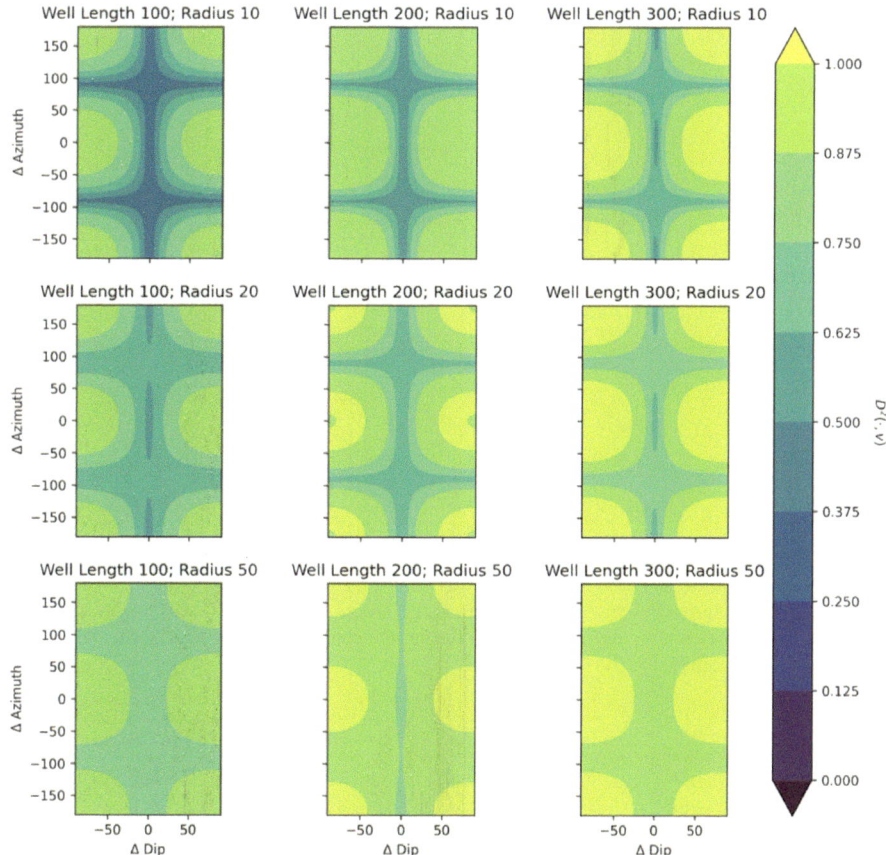

**Fig. 3** Illustration of dispersion variance within well drainage radius changing with azimuth and dip angle of the well trajectory deviating from major direction with well length = 100–300 m, well drainage radius = 10–50 m, as examples

change (%) between the group with high $D^2(\cdot, v)$ and the group with low $D^2(\cdot, v)$ to evaluate the sensitivity of the proposed metric for production. Figure 6 shows the cumulative oil production change under different Dykstra-Parsons coefficients and permeability magnitudes. Overall, the impact of dispersion variance is more sensitive when Dykstra-Parsons coefficient is high and the permeability mean is low. The sensitivity of the dispersion variance metric to cumulative production change per unit length increases as Dykstra-Parsons coefficient increasing.

When the permeability mean is low, i.e., close to the magnitude of tight oil/shale reservoir, the dispersion variance within well drainage radius is informative, indicating high and low cumulative production. When permeability mean is high, well drainage radius is no longer a fixed value near-wellbore anymore, extending to the whole reservoir. That explains why dispersion variance reflects less information for

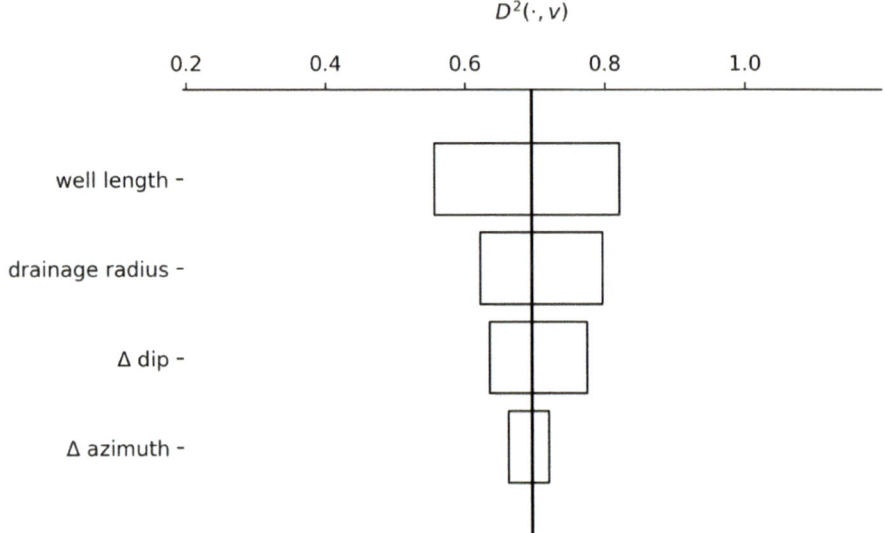

**Fig. 4** Tornado plot for the result of sensitivity analysis of the proposed spatial-engineered feature, standardized dispersion variance within well drainage radius, for different well length, drainage radius, well trajectories (i.e., dip and azimuth deviating from major direction of the geological spatial continuity model)

**Table 2** Base case and test case range of the sensitivity analysis for standardized dispersion variance within well drainage radius with different well length, drainage radius, well trajectories (i.e., dip and azimuth deviating from major direction of the geological spatial continuity model)

|                     | Base case | Test case range |
| ------------------- | --------- | --------------- |
| Well length (m)     | 218.75    | 50~500          |
| Drainage radius (m) | 21.88     | 5~50            |
| $\Delta$ dip ($^\circ$)       | 10        | 0~20            |
| $\Delta$ azimuth ($^\circ$)   | 45        | 0~180           |
| Well length (m)     | 218.75    | 50~500          |

**Fig. 5** Cumulative oil production expectation curve (solid line) and 95% confidence interval (dash line) grouped by high and low dispersion variance feature values with Dykstra-Parsons coefficient = 0.8, permeability mean = 0.08 mD

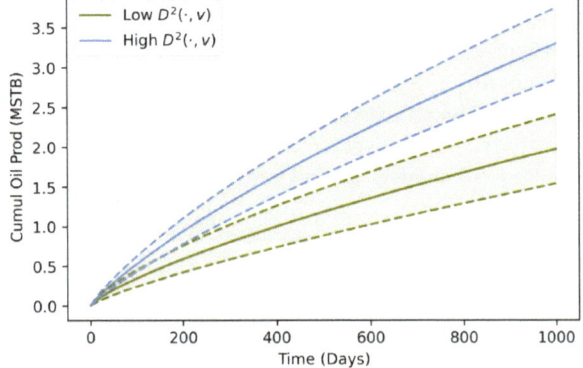

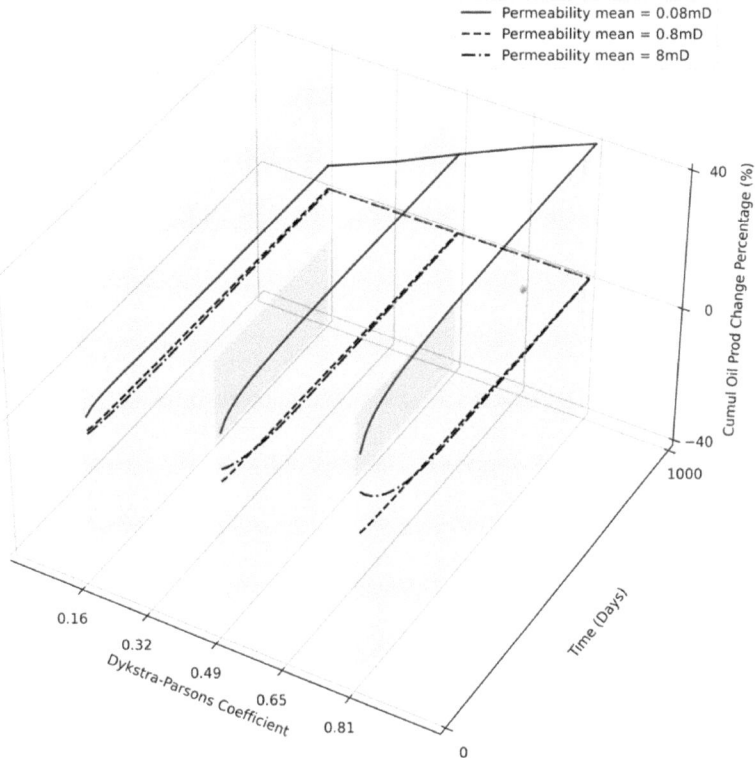

**Fig. 6** Cumulative oil production change between the group curves with high $D^2(\cdot, v)$ and the group with low $D^2(\cdot, v)$ under different Dykstra-Parsons coefficient and permeability magnitudes (solid: permeability mean $= 0.08$ mD; dash: permeability mean $= 0.8$ mD; dot: permeability mean $= 8$ mD)

production, as dispersion variance within well drainage radius now converges to the reservoir volume support size, which is equal to 1 in standardized scale.

## 4  Case Study

Based on the analysis from the previous section, we can infer the dispersion variance could be an informative predictor feature for tight oil or shale reservoirs. Therefore, we further demonstrate the dispersion variance as a spatial feature for the predictive machine learning model with a case study in the Kaybob field, Duvernay formation.

The Duvernay formation was deposited in a sub-equatorial epicontinental seaway in the late Devonian, Frasnian time, this corresponds to the maximum transgression of this late Devonian sea into the western Canadian craton. The shale is deposited in the paleo-lows within the confines of the surrounding Leduc reefs in a slope and basin

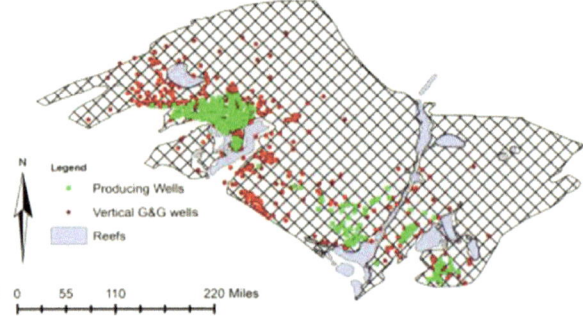

**Fig. 7** Duvernay formation map with vertical G&G wells (red) and horizontal producers (green)

environment. The results in a series of sub-basins deposits from the west shale to east shale basins. The depth ranges from 2000–3700 m and the formation produces across the oil, condensate and gas windows. The greater Kaybob area locates in the West shale basin. The formation is at greatest thickness in the Kaybob area thinning to the east. The mineralogy also changes from West to east. The Kaybob area is a more silica-rich shale passing into the less quartz-rich, higher clay and higher carbonate content East shale basin [27, 28]. The West shale basin, Kaybob area is the most developed, where the majority of Duvernay production comes from (see Fig. 7).

For this case study, we use 110 horizontal wells with features and response listed in Table 3. We choose to use production per unit length as the response feature to remove the direct impact from well length and use barrel of oil equivalent (BOE) as a convenient summarization of production response for oil, gas and condensate volumes in the same units. The dispersion variance within well drainage radius is calculated based on the porosity variogram model and the maximum drainage radius for each well is approximated based on well spacing. A non-parametric conditional expectation plot of production given the dispersion variance is shown in Fig. 8. To further investigate the impact of our proposed spatial-engineered feature on a predictive machine learning model, we test the dispersion variance feature in random forest and gradient boosting models. By grid search and k-fold validation where k = 5, we find the optimal hyperparameters and the average of the metrics calculated from the k-fold cross-validation testing sets of the optimal model with standard deviation shown in Fig. 9, where the optimal models in Fig. 9a use all the features in Table 3 while the optimal models in Fig. 9b exclude the dispersion variance feature. Including the proposed spatial feature reduces the mean absolute error (MAE) and rooted mean squared error (RMSE) for random forest. While for gradient boosting model, adding the spatial-engineered feature only reduces MAE. Since both random forest and gradient boosting are stochastic models, we iterate over 100 realizations with different random seeds to check if the performance is stable. The relative difference of metrics (MAE, RMSE) is defined as the metric for the model with the spatial-engineered feature minus that without the spatial-engineered feature, over the metric value with the spatial-engineered feature. The relative difference distribution is shown in Fig. 10. In the majority of cases, including the proposed spatial-engineered feature improves the model performance, judging from a smaller MAE for both random forest and gradient boosting. This work demonstrates that our proposed dispersion variance feature may, in some cases, improve predictions and should be considered as a new spatially informed feature in building predictive models. Demonstrating the rank of feature importance of dispersion variance for predicting production relative to all other possible geological and engineering parameters would require a much more comprehensive study and is not the goal of this work.

**Table 3** Summary of available data for the predictive machine learning model

|  | Type | Name | Field unit | SI unit |
|---|---|---|---|---|
| Response | Production | First 12-month production | BOE/1000ft | $m^3$/305 m |
| Predictor features | Completion | Fracture stages | count | count |
|  |  | Average stage spacing | ft | m |
|  |  | Total fluid pumped | bbl | $m^3$ |
|  |  | Amount of proppant | lbs | kg |
|  |  | Proppant intensity | lbs/ft | kg/m |
|  |  | Fluid intensity | lbs/ft | kg/m |
|  | Petrophysics | Water saturation | fraction | fraction |
|  |  | Porosity | fraction | fraction |
|  | Spatial feature | Dispersion variance | fraction | fraction |

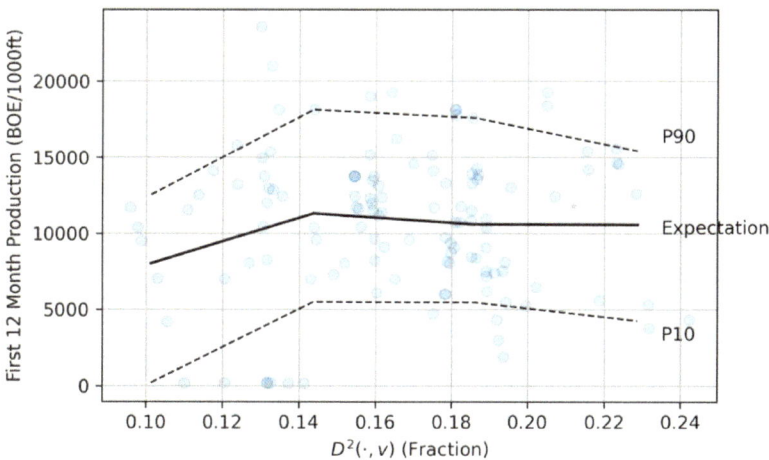

**Fig. 8** Conditional P10 (dash line), expectation (solid line), and P90 (dash line) of first 12-month cumulative production to dispersion variance within well drainage radius over 110 wells (scatter)

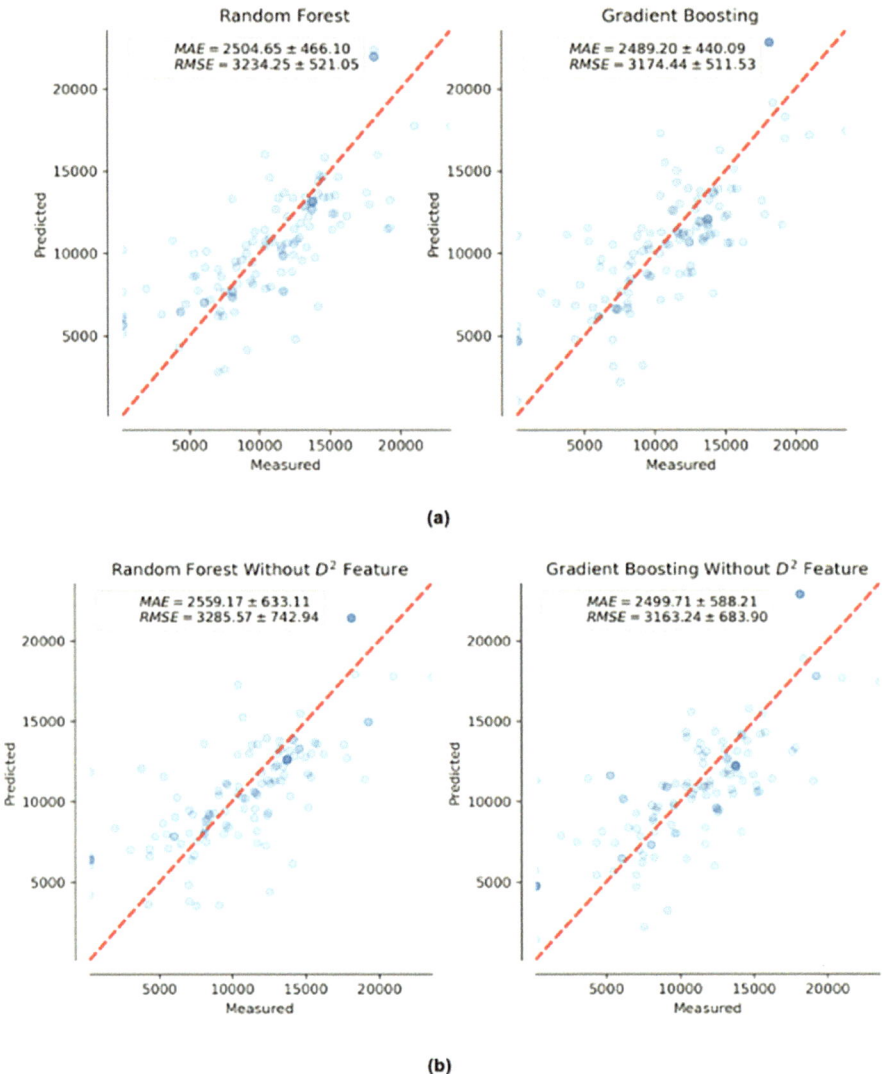

**Fig. 9** Measured and predicted first 12-month cumulative production (scatter) along the 45-degree line (dash line) with average metrics (mean absolute error and root mean squared error) $\pm$ the corresponding standard deviation via k-fold validation, k = 5 (**a**) using random forest and gradient boosting with all the features in Table 3 (**b**) without dispersion variance feature

## 5  Conclusion

The proposed spatial-engineered heterogeneity feature, well dispersion variance, integrates the impact of spatial continuity and data volume support size and is computationally efficient to calculate. Dispersion variance is sensitive to various spatial

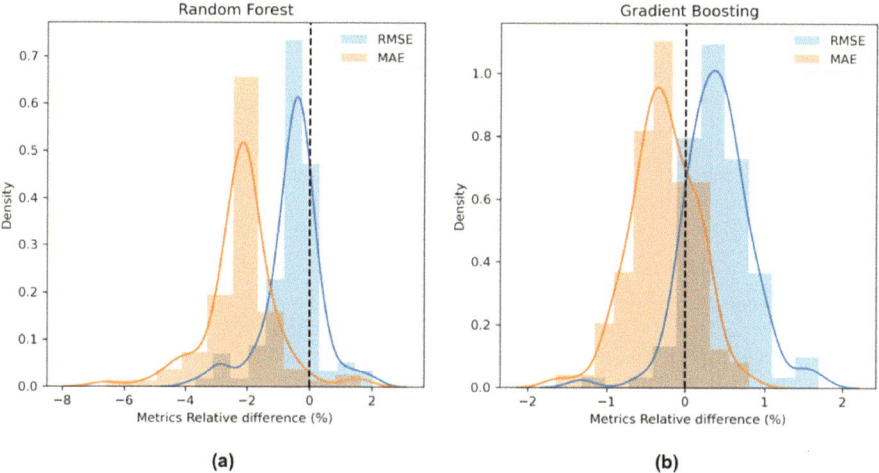

**Fig. 10** Relative difference distribution of the metrics, mean absolute error and root mean squared error, over 100 realizations with different random seeds for (**a**) random forest models and (**b**) gradient boosting models

factors, such as well trajectories with respect to major direction, and to the response feature for flow through porous media. Integrating spatial, scale information into a single, spatially aware feature also helps to reduce dimensionality for predictive machine learning models and improve the model performance. We demonstrate the spatial-engineered feature could be specifically useful for unconventional or tight oil reservoirs. We suggest the augmentation of predictive machine learning models with the proposed spatially engineered feature for improving subsurface resource prediction.

# References

1. Zheng, A., Casari, A.: Feature engineering for machine learning: principles and techniques for data scientists. O'Reilly Media, Inc. (2018)
2. Ozdemir, S., Susarla, D.: Feature engineering made easy: Identify unique features from your dataset in order to build powerful machine learning systems. Packt Publishing Ltd. (2018)
3. Amaefule, J.O., Altunbay, M., Djebbar, T., Kersey, D.G., Keelan, D.K.: Enhanced Reservoir description: using core and log data to identify hydraulic (flow) units and predict permeability in uncored intervals/wells. In: SPE Annual Technical Conference and Exhibition, Houston, Texas (1993)
4. Goodfellow, I., Bengio, Y., Courville, A., Bengio, Y.: Deep learning. MIT press, Cambridge (2016)
5. Liu, W., Ikonnikova, S., Hamlin, H.S., Sivila, L., Pyrcz, M.J.: Demonstration and mitigation of spatial sampling bias for machine-learning predictions. SPE Reservoir Eval. Eng. **24**(01), 262–274 (2021)
6. Liu, W., Pyrcz, M.J.: A spatial correlation-based anomaly detection method for subsurface modeling. Math. Geosci. **53**(5), 809–822 (2021)

7. Salazar, J.J., Garland, L., Ochoa, J., Pyrcz, M.J.: Fair train-test split in machine learning: mitigating spatial autocorrelation for improved prediction accuracy. J. Pet. Sci. Eng. **109885**. https://doi.org/10.1016/j.petrol.2021.109885 (2021)
8. Salazar, J.J., Pyrcz, M.J.: Geostatistical significance of differences for spatial subsurface phenomenon. J. Pet. Sci. Eng. **203** (2021)
9. Ahmed, T.: Reservoir Engineering Handbook, 5th edn. Gulf Professional Publishing (2019)
10. Dykstra, H., Parsons, R.L.: The prediction of oil recovery by water flood, secondary recovery of oil in the United States. Principle and Practice (1950)
11. Schmalz, J.P., Rahme, H.D.: The variation of waterflood performance with variation in permeability profile. Prod. Mon. **15**(9), 9–12 (1950)
12. Lake, L.W., Jensen, J.L.: A review of heterogeneity measures used in reservoir characterization. In Situ **15**(4), 409–440 (1991)
13. Polasek, T.L., Hutchinson, C.A.: Characterization of non-uniformities with a sandstone reservoir from a fluid mechanics standpoint. In: Proceedings of the Seventh World Petroleum Congress, Elsevier Pub. Co. Ltd., vol. 2, pp. 397–407 (1967)
14. Alpay, O.A.: A Practical approach to defining reservoir heterogeneity. J. Pet. Technol. 841–848 (1972)
15. Hassibi, M., Ershaghi, I., Aminzadeh, F.: High resolution reservoir heterogeneity characterization using recognition technology. In: Developments in Petroleum Science, vol. 51, pp. 289–307. Elsevier (2013)
16. Martinius, A.W., Fustic, M., Garner, D.L., Jablonski, B.V.J., Strobl, R.S., MacEachern, J.A., Dashtgard, S.E.: Reservoir characterization and multiscale heterogeneity modeling of inclined heterolithic strata for bitumen-production forecasting, McMurray Formation, Corner, Alberta, Canada. Mar. Pet. Geol. **82**, 336–361 (2017)
17. Zhang, L., Li, B., Jiang, S., Xiao, D., Lu, S., Zhang, Y., Gong, C., Chen, L.: Heterogeneity characterization of the lower Silurian Longmaxi marine shale in the Pengshui area, South China. Int. J. Coal Geol. **195**, 250–266 (2018)
18. Journel, A.G., Huijbregts, C.J.: Mining geostatistics. Academic Press, New York (1978)
19. Pyrcz, M.J., Deutsch, C.V.: Geostatistical reservoir modeling, 2nd edn. Oxford University Press, New York (2014)
20. Frykman, P., Deutsch, C.: Practical application of geostatistical scaling laws for data integration. Petrophysics. **43**, 153–171 (2002)
21. Lake, L.W., Srinivasan, S.: Statistical scale-up of reservoir properties: concepts and applications. J. Petrol. Sci. Eng. **44**, 27–39 (2004)
22. Pyrcz, M.J.: Data analytics and geostatistical workflows for modeling uncertainty in unconventional reservoirs. Bull. Can. Pet. Geol. **67**(4), 273–282 (2019)
23. Okabe, A., Boots, B., Sugihara, K., Chiu, S.N.: Spatial tessellations: concepts and applications of voronoi diagrams, 2nd edn. Wiley, New York (2000)
24. Pyrcz, M.J., Deutsch, C.V.: Declustering and debiasing. Technical report, Center for Computational Geostatistics. http://www.gaa.org.au/pdf/DeclusterDebias-CCG.pdf (2003)
25. Deutsch, C.V., Journel, A.G.: GSLIB Geostatistical software library and user's guide, 2nd edn. Oxford University Press, New York City (1997)
26. Pyrcz, M.J., Jo. H., Kupenko, A., Liu, W., Gigliotti, A.E., Salomaki, T., Santos, J.: GeostatsPy Python Package, PyPI, Python Package Index. https://pypi.org/project/geostatspy/ (2021)
27. Switzer, S.B., Holland, W.G., Christie, D.S., Graf, G.C., Hedinger, A., McAuley, R., Wierzbicki, R., Packard, J.J.: Devonian Woodbend-Winterburn strata of the Western Canada sedimentary basin. Chapter 12. In: Mossop, G.D., Shetsen, I. (eds.) Geological Atlas of the Western Canada sedimentary basin. Canadian Society of Petroleum Geologists and Alberta Research Council, p. 165–202 (1994)
28. Rokosh, C.D., Lyster, S., Anderson, S.D.A., Beaton, A.P., Berhane, H., Brazzoni, T., Chen, D., Cheng, Y., Mack, T., Pana C., Pawlowicz, J.G.: Summary of Alberta's Shale- and Siltstone-hosted hydrocarbons. In: ERCB/AGS Open File Report 2012-06, p. 327 (2012)

# Iterative Gaussianisation for Multivariate Transformation

A. Cook, O. Rondon, J. Graindorge, and G. Booth

**Abstract** Multivariate conditional simulations can be reduced to a set of independent univariate simulations through multivariate Gaussian transformation of the drill hole data to independent Gaussian factors. These simulations are then back transformed to obtain simulated results that exhibit the multivariate relationships observed in the input drill hole data. Several transformation techniques are cited in geostatistical literature for multivariate transformation. However, only two can effectively simulate high dimensional drill hole data with complex non-linear features: Flow Anamorphosis (FA) and Projection Pursuit Multivariate Transformation (PPMT). This paper presents an alternative iterative multivariate Gaussian transformation (IG) along with a multivariate simulation case study of a large Nickel deposit. Our findings show that IG is computationally faster than FA and PPMT which makes the technique more appealing for most practical and time-sensitive applications.

**Keywords** Multivariate Gaussian transformation · Conditional simulation · Projection pursuit

## 1 Introduction

Traditional univariate conditional simulation techniques transform the data to Gaussian space via quantile–quantile transformation [1] where the simulation proceeds and the results back-transformed to input data space via the corresponding back-transformation. Conventional multivariate simulation techniques like Cosimulation

A. Cook · O. Rondon (✉)
Snowden Optiro, Level 19, 140 St Georges Terrace, Perth 6000, Australia
e-mail: oscar.rondon@snowdenoptiro.com

J. Graindorge
Fortescue Metals Group, Level 2/87 Adelaide Terrace, East Perth, WA 6004, Australia

G. Booth
Ambatovy Minerals, Tranofitaratra Rue Ravoninahitriniarivo Antananarivo, 101 Antananarivo, Madagascar

© The Author(s) 2023
S. A. Avalos Sotomayor et al. (eds.), *Geostatistics Toronto 2021*, Springer Proceedings in Earth and Environmental Sciences, https://doi.org/10.1007/978-3-031-19845-8_2

[2], PCA [3, 4] and MAF [4] transform each attribute separately using the quantile–quantile approach and the simulation proceeds by assuming the transformed data have multivariate Gaussian distribution. This assumption is not realistic because having marginal Gaussian distributions does not necessarily ensure the transformed data has multivariate Gaussian distribution. This is no longer an issue for truly multivariate techniques like stepwise conditional transformation (SCT) by Leuangthong et al. [5], Projection Pursuit Multivariate Transformation (PPMT) by Barnett et al. [6] and Flow Anamorphosis (FA) by van den Boogaart [7].

SCT workflow is cumbersome and often difficult to apply with increasing number of attributes [8] which has limited its used in practical applications. FA has the ability to handle a large number of attributes but depends on two parameters, both of which require significant fine-tunning to ensure convergence to a multivariate Gaussian distribution. Further to this, in its current form, FA is impractical for large data sets due to the significant computational processing time required for processing the data. Conversely, PPMT's lower overall processing time and minimal tunning parameters have made the technique more appealing for most practical.

This paper presents a multivariate simulation case study of a large Nickel deposit using an iterative multivariate Gaussian transformation developed by Laparra et al. [9] for image processing. The technique does not require any tunning parameters, nor does it need to search for the most non-Gaussian projection, which is key to PPMT. Furthermore, its direct and back transformations are faster and convergence to a standard multivariate Gaussian distribution is proven. This position IG as a technique that may supplant PPMT and FA for time-sensitive applications.

## 2  Iterative Multivariate Gaussianisation

IG is simply a sequential application of univariate marginal Gaussianisation using the quantile–quantile approach followed by a rotation using an orthonormal transformation [9]. An important aspect of IG is that the type of rotation is not critical because the algorithm convergence is proven for any orthonormal transformation.

Let $X^{(0)}$ be the multivariate input data and $\Psi\left(X^{(0)}\right)$ the marginal Gaussianisation of each dimension of $X^{(0)}$. Then the iterative process is defined as

$$X^{(k+1)} = R_k \Psi\left(X^{(k)}\right) \tag{1}$$

where $R_k$ is a generic rotation matrix for $\Psi\left(X^{(k)}\right)$ [9]. A simple choice is set $R_k$ to be the matrix of eigenvectors of $\Psi\left(X^{(k)}\right)$. This provides an easily programmable closed-form for the direct and back-transformations. Furthermore, Srivastava's skewness and kurtosis multivariate normality test [10] can be seamlessly integrated to derive appropriate stopping conditions. As pointed out by Laparra et al. [9], using the eigenvectors also guarantees the algorithm convergence except for the case of a multivariate input data having all its univariate marginal distributions equal to the standard Gaussian distribution. This case can rarely be found in real data sets.

# 3   Nickel Laterite Case Study

## 3.1   Overview

This case study presents the validation results for a multivariate conditional simulation utilising IG for a large nickel laterite deposit. The simulation was generated as part of a Drillhole Spacing Analysis (DSA) with the aim of quantitatively assessing the economic cost vs risk at varying sample densities, based on the quality of the grade estimation and potential for misclassification. Due to the correlated nature of the input variables (Ni, Co, MgO, $SiO_2$ and $Al_2O_3$), the simulation required a multivariate approach to ensure that the correlations were reproduced and maintained.

The nickel laterite study area is approximately 2 $km^2$ and is informed by 97,682 samples with variable spacing as shown in Fig. 1. For each sample, multielement data is available from which 5 key elements have been analysed, due to their economic interest to the mine operator. A single lithological domain was identified as the target of the study and warranted basic unfolding techniques to minimise variations in mineralisation orientation.

**Fig. 1**  Plan section of the case study area showing all sample locations

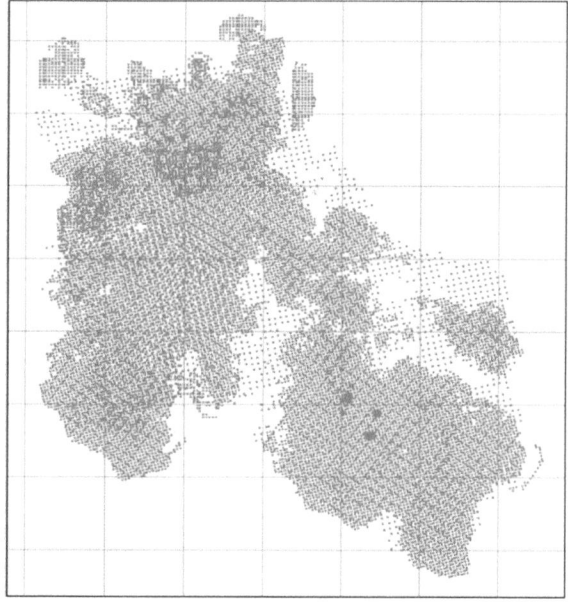

250m

## 3.2 Workflow

The high-level steps followed to generate the multivariate simulations from the input data for the study area is outlined below.

1. Perform multivariate and compositional exploratory data analysis to confirm variable correlations and validate the composition.
2. Transform coordinates to unfolded space using basic z-only transform for flattening.
3. Perform compositional transformation using an appropriate log-ratio technique In this case, the additive log-ratio (ALR) transformation was used.
4. Perform multivariate normal transformation using IG.
5. Validate statistical properties and spatial decorrelation of independent Gaussian factors.
6. Simulate in Gaussian space using sequential gaussian.
7. Back-transform assays to compositional space, then to raw space.
8. Refold simulations to raw coordinate space.
9. Validate simulation results with input data.

## 3.3 Multivariate Transformation and Simulation

Multivariate techniques are suitable for element compositions which exhibit correlations and form a composition, or sub-composition. Bivariate analysis of the chosen input variables demonstrated complex non-linear relationships which must be preserved in the simulation results, as shown in Fig. 2.

**Fig. 2** Hexbin plot showing the input correlation between $SiO_2$ and $Al_2O_3$

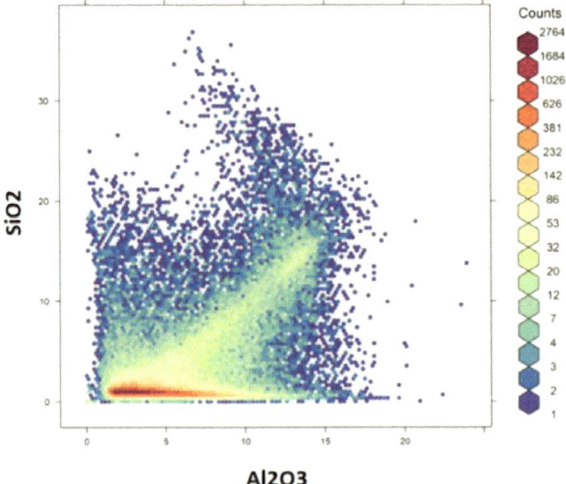

When the variables under study form a sub-composition, i.e. several variables jointly describe the relative weight with respect to a whole, a form of completing is required to ensure the constant sum constraint, which is the case for the multivariate simulation of Ni, Co, MgO, $SiO_2$ and $Al_2O_3$. This is achieved by defining a filler variable which represents all other variables not being considered that make up the remainder of the sample composition. In this case, additive log-ratio (ALR) was selected as the transformation method to be used to unconstraint the sub-composition formed by Ni, Co, MgO, $SiO_2$ and $Al_2O_3$. Furthermore, ALR is simple and suited to work with conditional simulations [11].

The IG method was used to further transform the ALR data into equivalent independent factors with multivariate standard Gaussian distribution. IG was chosen because PPMT produced artifacts in the resulting factors regardless of the number of iterations used during the transformation. An example is shown in Fig. 3. The scatterplot between two factors computed using the PPMT transform exhibit linear stripes that do not correspond to the expected scatterplot between independent Gaussian attributes with multivariate Gaussian distribution. Furthermore, IG's simplicity and reasonably low runtime compared to PPMT and FA were considered major benefits.

Although IG theoretical properties ensure that factors have multivariate standard Gaussian distribution, it is not possible to know in advanced how many iterations are required to achieve convergence. In this study, 60 iterations were used but as shown in Sect. 3.4 much less iterations could have been used. Figure 4 shows all scatterplots between the derived four factors.

The degree of spatial correlation between factors was assessed visually by computing omnidirectional cross variograms for up to 400 m (Fig. 5). The results show that spatial correlation between factors can be considered negligible with an absolute maximum value of 0.14.

Factors were simulated using the Sequential Gaussian Simulation [1]. The simulations informed a 2.5 mN × 2.5 mE × 2 mRL grid of nodes and were considered

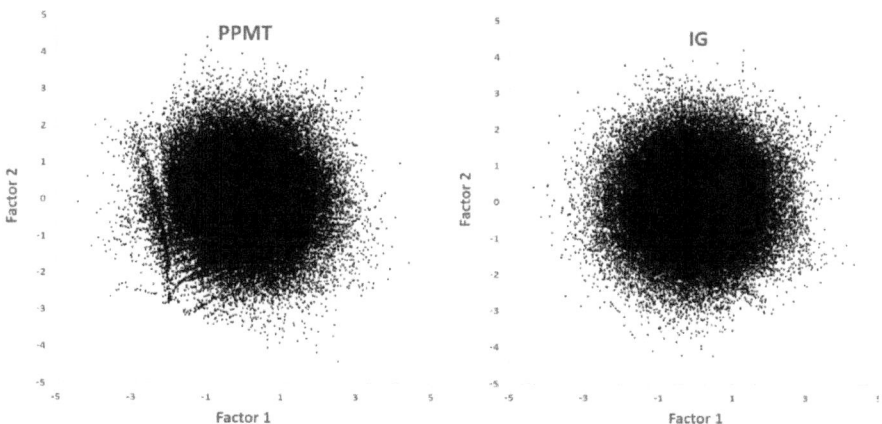

**Fig. 3**  Scatterplots between factors for PPMT (left) and IG (right)

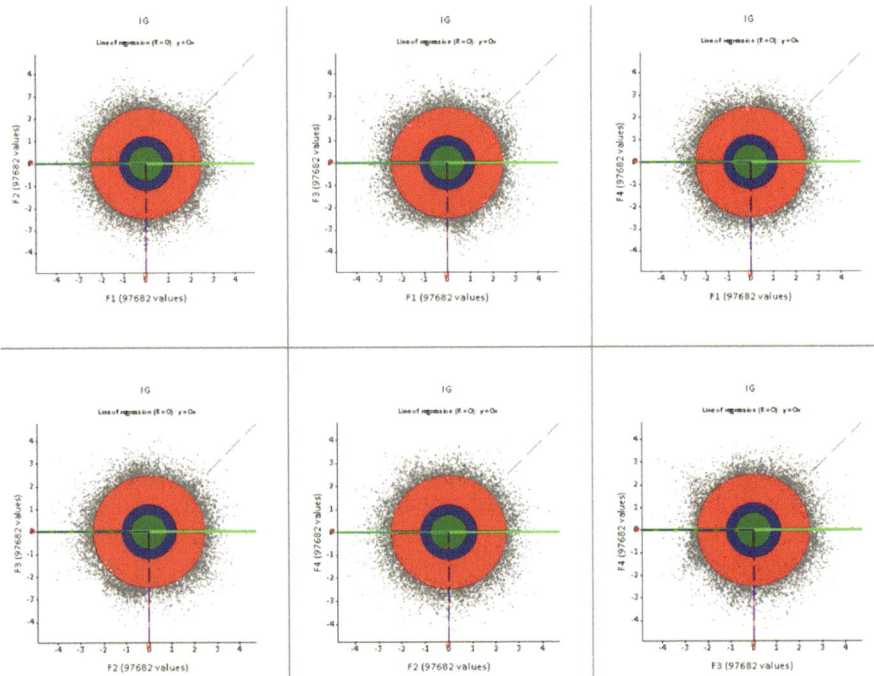

**Fig. 4** Scatterplots between factors derived using the IG transformation. Data is shown in grey along with the confidence ellipses for the 95th (red), 50th (blue) and 15th (green) percentiles according to a standard multivariate Gaussian distribution

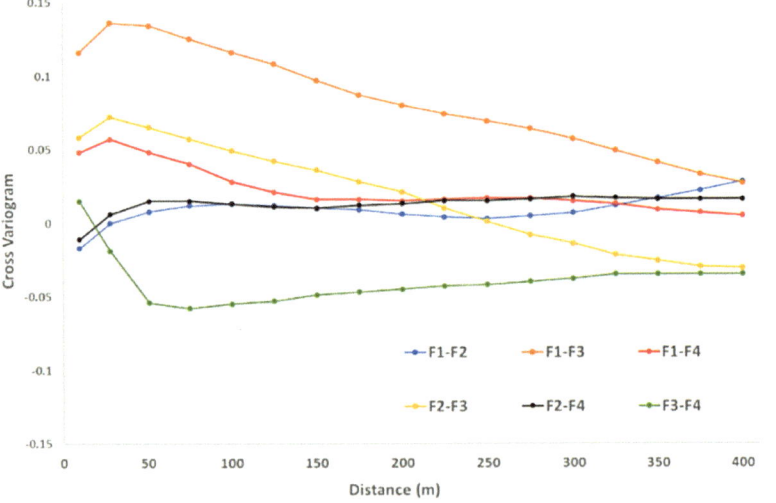

**Fig. 5** Omnidirectional cross variograms between factors derived using the IG approach

the point support simulations. Scatterplots were generated for each of the element pairs to ensure that the data correlations were reproduced in the simulation results. Figure 6 demonstrates that the IG technique has successfully maintained the complex non-linear correlations in the input data.

Trend plots of the average naïve and de-clustered composites and a single simulation of nickel are presented in Fig. 7. The simulated grades in the trend plots demonstrate minor deviation from the input drillhole data; however, this is primarily influenced by local variability introduced by the simulation process and irregular sample distribution compared to the cell volumes. Globally, the difference amounts to an 8–10% difference; however, majority of simulated cells generally exhibit much lower differences. Visual inspection across the model shows good correlation to immediately surrounding samples and the overall trends of the grades across the deposit.

Figure 8 illustrates east–west profiles for three realisations at point support (2.5 mE × 2.5 mN × 2 mRL spacing) of simulated nickel mineralisation, with the conditioning drillhole data, in unfolded space. The simulations show good reproduction of the input data and reflect the mineralization trends and continuity that were evident in

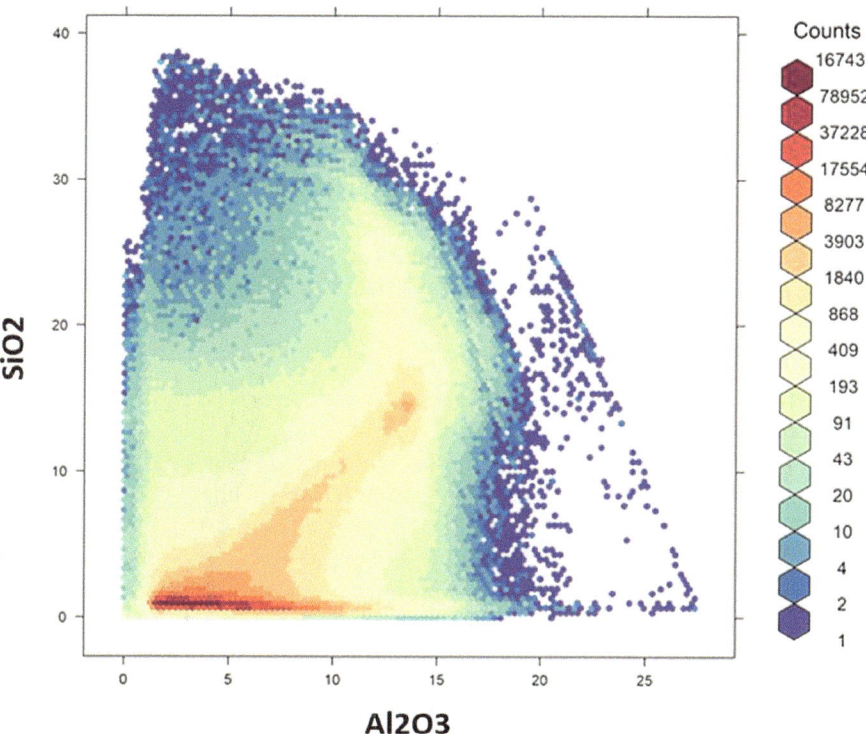

**Fig. 6** Hexbin plot showing the correlation between $SiO_2$ and $Al_2O_3$ for a single, randomly selected realization

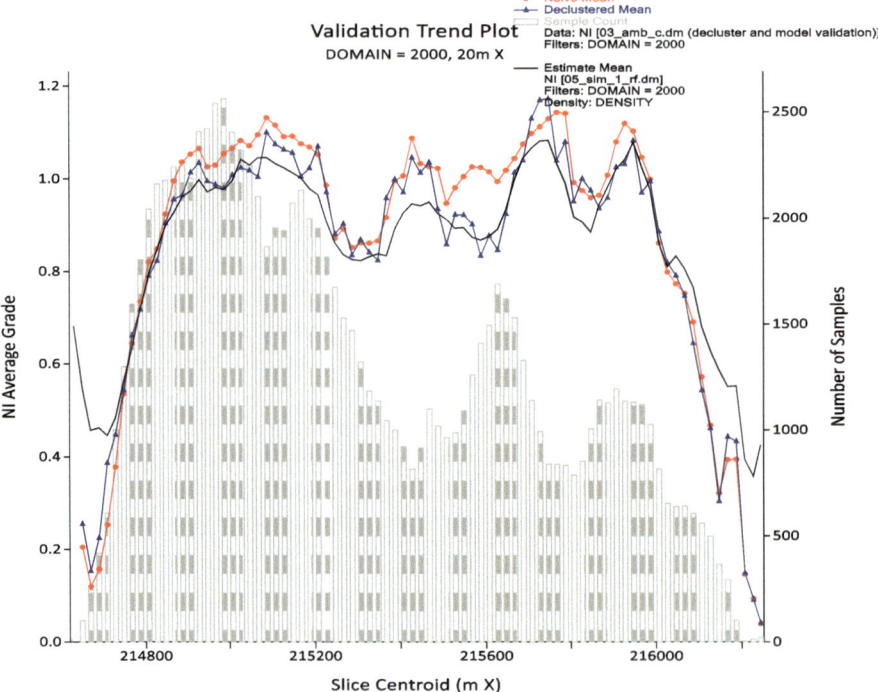

**Fig. 7** Trend plot for nickel comparing the naïve (red) and declustered (blue) sample data to a randomly selected simulation (black)

the spatial analysis. Additionally, there is good alignment between the simulations with the greatest variability occurring where data is sparser, and the grade data is less continuous.

The conclusions from the validation of the simulations, are:

- Visual comparison of the simulated grades and the corresponding drillhole grades showed reasonable correlation.
- A comparison of the global drillhole and simulated domain grades for Ni, Co, $SiO_2$ and $Al_2O_3$ shows that the mean grades of the simulations were typically within 5%.
- Comparison of the variance of the input composite data against the simulations shows that the simulations adequately reproduce the variance of the input data.
- Analysis of the correlation coefficients between Ni, Co, MgO, $SiO_2$ and $Al_2O_3$ for each deposit shows that the correlations of the input composite data are reproduced in the simulated grades. Furthermore, the compositional closure is preserved, as demonstrated in Fig. 6.
- The input data contains some outlying correlations which the simulations attempt to reproduce and may appear to be artefacts in the scatterplots. These samples are considered to be real and therefore included in the dataset without no top cutting

Nickel

| | | | | | |
|---|---|---|---|---|---|
| [ABSENT] | [0.2,0.45] | [0.65,0.8] | [0.9,1] | [1.1,1.25] | [1.4,CEILING] |
| [FLOOR,0.2] | [0.45,0.65] | [0.8,0.9] | [1,1.1] | [1.25,1.4] | |

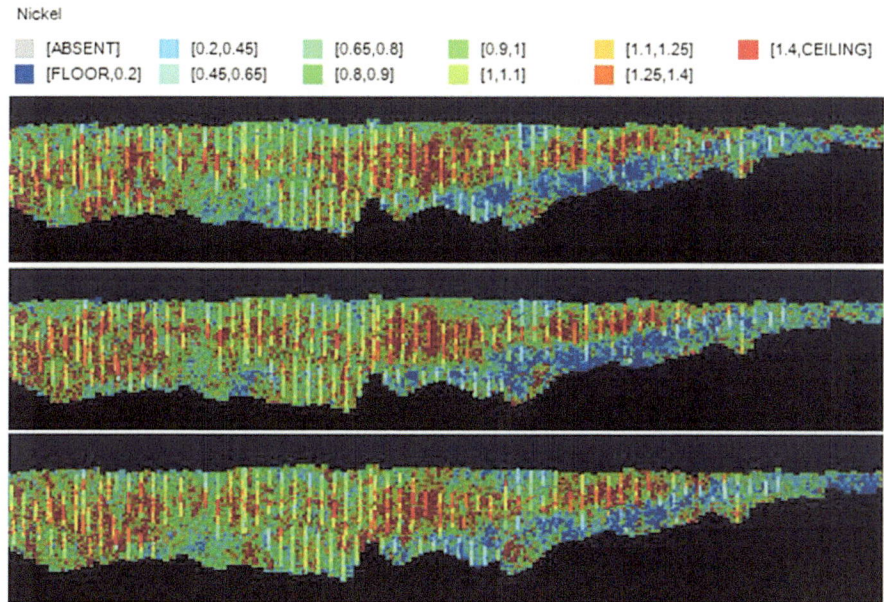

**Fig. 8** Northeast-southwest cross-section showing three point support (2.5 m × 2.5 m × 2 m) realisations of nickel mineralisation, with conditioning data, in unfolded space

or filtering so that the variability of all aspects of the dataset were reproduced. The number of records which make up these outlying correlations amount to less than 1% of the total dataset.

- Except for poorly sampled regions, the grade trend plots show a good correlation between simulated and drillhole grades.

The simulations are therefore considered a suitable representation input characteristics observed in the drill hole data.

## 3.4 Benchmarking

PPMT and IG were compared by analysing the run time as function of increasing number of samples and by testing the rapidness of the convergence to a standard multivariate Gaussian distribution using the Energy test [12]. Flow anamorphosis was not considered during the benchmark due to the long run time required to get the results.

Two run-time tests were conducted to directly compare the total processing time required for each of the IG and PPMT methods. For sample numbers between 10,000 and 50,000 there is a significant time saving when using IG of approximately 90% with an average of 953 samples processed per second compared to PPMT's

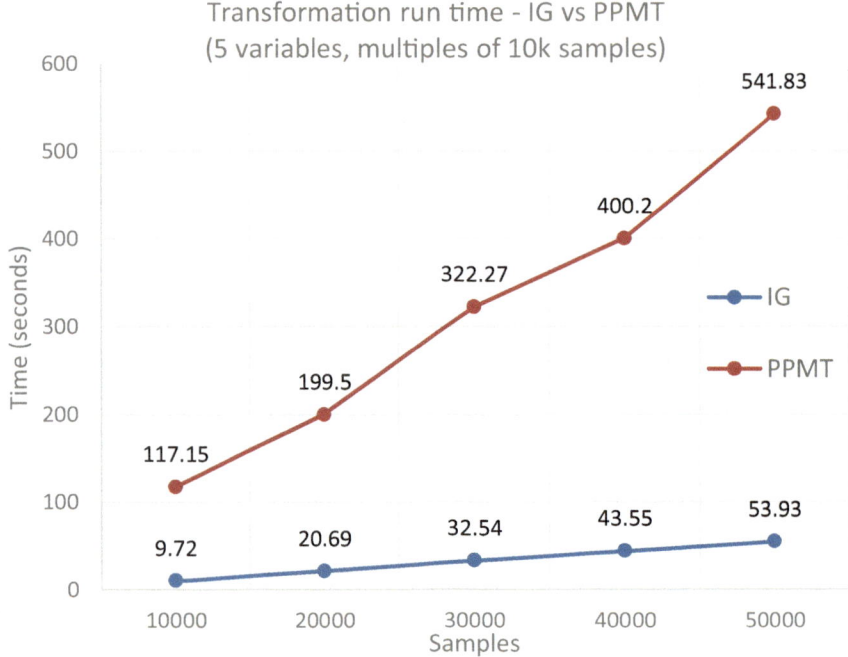

**Fig. 9** Comparison of total run-time for both IG and PPMT for multiples of 10,000 samples

94 samples per second (Fig. 9). Further testing for increasing numbers of samples between 10,000 and 10,000,000 samples indicate that the ratio further increases with greater populations (Fig. 10). In addition, PPMT was unable to complete the ten million sample run in the test environment analysed.

Results for the Energy test are shown in Fig. 11. For each iteration, the test was carried out using a 95% confidence level and the resultant P-value reported. The results show that IG requires a fraction of the iterations used by PPMT to converge to a standard multivariate Gaussian distribution.

## 3.5 Artifacts

As with many techniques, a core difficulty is the reproduction of under-represented features and extreme values. An artifact in the data is considered to be where the technique fails to reproduce geological features and relationships in a manner that would be expected in the geological setting. Comparison of IG, PPMT and FA techniques and their ability to minimise artifacts in the presence of extreme values are illustrated in Fig. 12, Fig. 13 and Fig. 14 respectively. Between the three techniques,

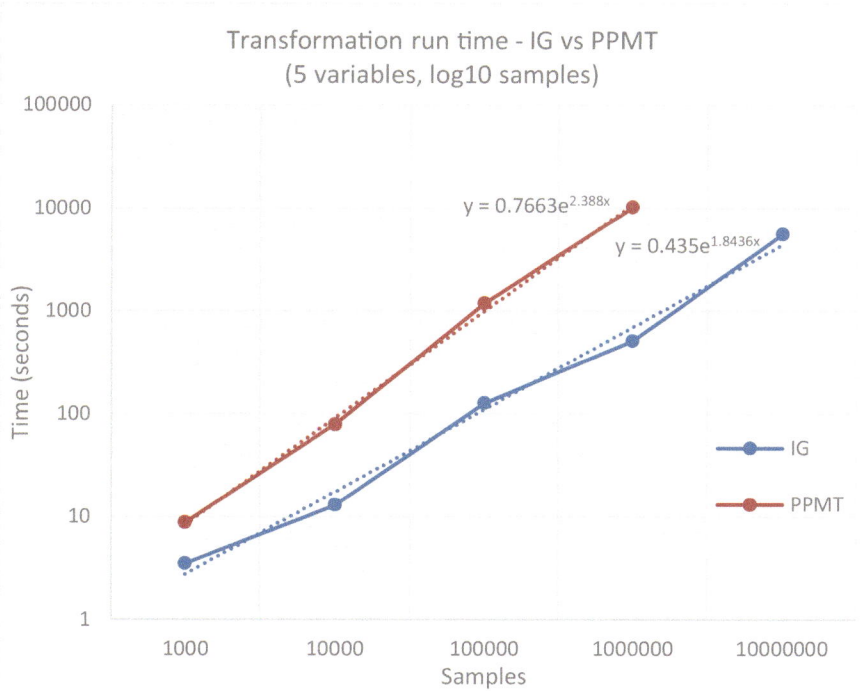

**Fig. 10** Comparison of total run-time for both IG and PPMT for log10 sample quantities

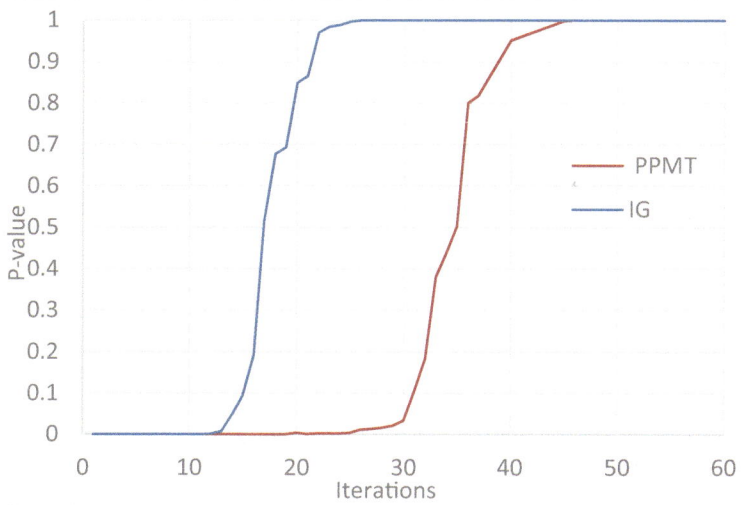

**Fig. 11** Comparison of calculated p-value for both IG and PPMT by iteration number

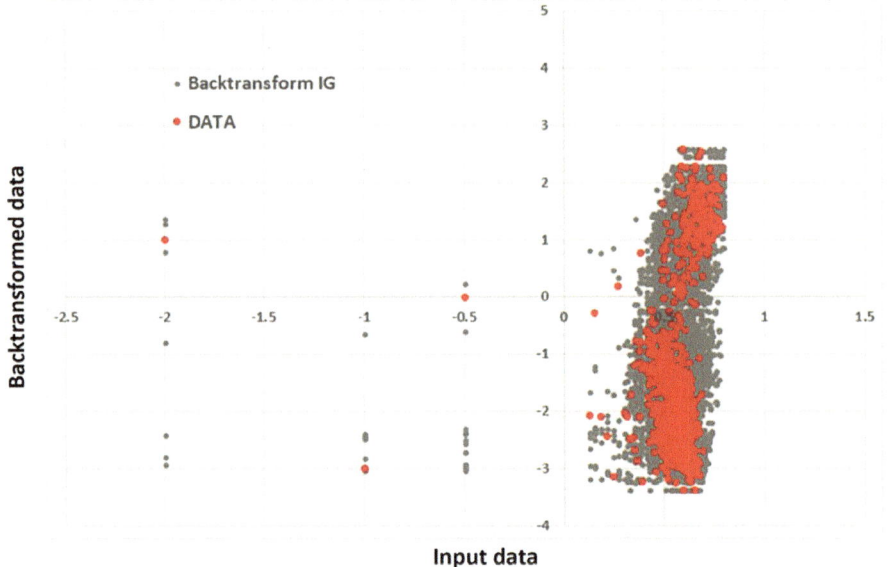

**Fig. 12** Comparison of input data versus backtransformed values for the IG method

only FA significantly minimises the effect of extreme values. PPMT provides a few key benefits when compared to the IG results, while retaining some issues with values in regions uninformed by the input data.

Comparison of the drillhole data correlations with the simulation results are shown in Fig. 15. These graphs highlight areas where artifacts are most significant due to the values being extreme for the dataset and the compositional transformation ensuring closure. While these features are not typical of a raw geological dataset, the relationships are acceptable within the context of the deposit. In addition, these artifacts are generally pervasive where gaps in the relationships occur and could be improved through additional sampling if they were considered material to the interpretation of the results.

## 4   Conclusions

The validation work demonstrates that the simulations generated using compositional and iterative Gaussianisation techniques are valid and accurately represent the input data. In addition to the requirement for a valid technique, many mine production settings require further criteria for long-term uptake of new mathematical techniques and must:

1. Produce results within a timely manner to meet time-sensitive targets for large populations of samples.

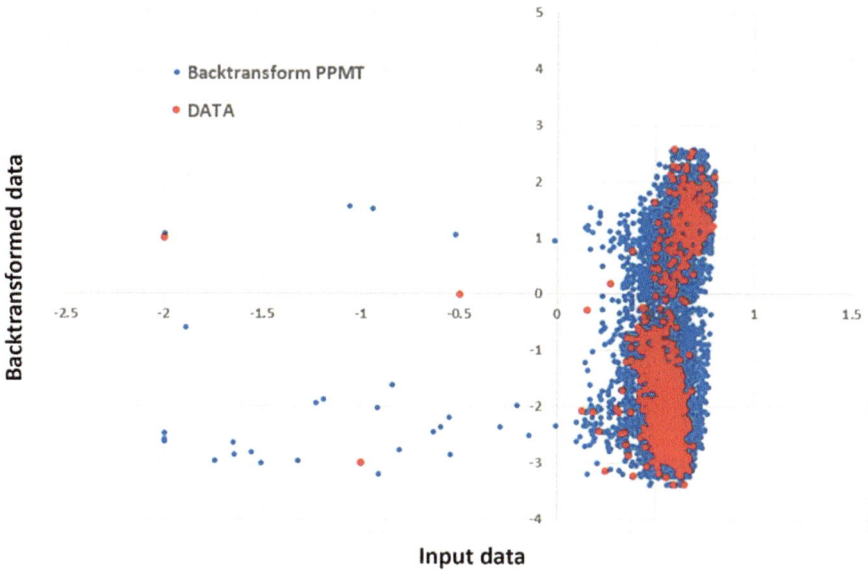

**Fig. 13** Comparison of input data versus backtransformed values for the PPMT method

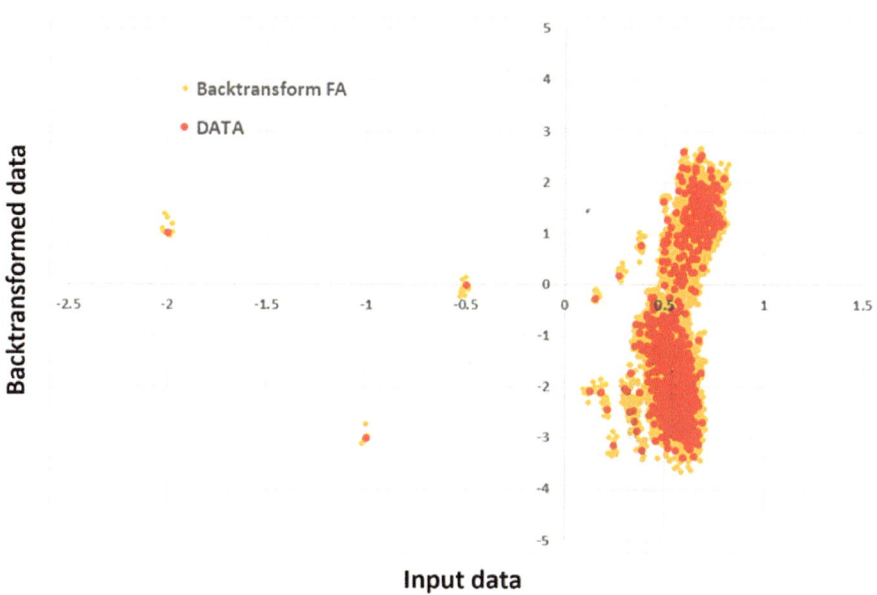

**Fig. 14** Comparison of input data versus backtransformed values for the FA method

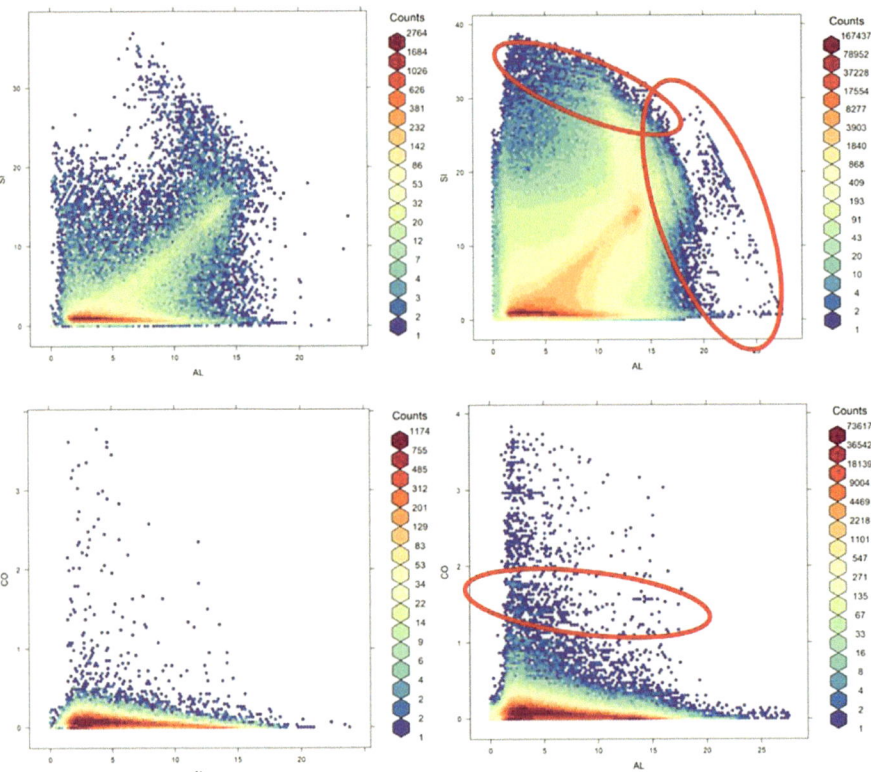

**Fig. 15** Comparison of input data (left) correlations compared to simulation results (right) for $SiO_2$–$Al_2O_3$ (top) and $Co$–$Al_2O_3$ with artifacts highlighted

2. Be usable in multiple settings, on a range of compositions with a low failure-rate.
3. Be easy to understand and utilise, as well as being openly available for the general resource estimator.

IG meets these criteria as the convergence to a gaussian distribution is always guaranteed, the matrices are always invertible and the technique is fast and simple. These benefits make the technique highly practical for the mining industry where time is precious and datasets exhibit complex relationships.

# References

1. Deutsch, C.V., Journel, A.G.: GSLIB: Geostatistical Software Library and User's Guide. Oxford University Press, Oxford (1998)
2. Chiles, J.P., Delfiner, P.: Modeling Spatial Uncertainty, 2nd edn. Wiley, New York (2012)
3. Bandarian, E., Bloom, L., Mueller, U.: Transformation methods for multivariate geostatistical simulations. In: Proceedings of the IAMG 06 XI-th International Congress for Mathematical Geology (2006)
4. Rondon, O.: Teaching aid: minimum/maximum autocorrelation factors for joint simulation of attributes. Math. Geosci. **44**(4), 469–504 (2012)
5. Leuangthong, O., Deutsch, C.V.: Stepwise conditional transformation for simulation of multiple variables. Math. Geol. **35**, 155–173 (2003)
6. Barnett, R.M., Manchuk, J.G., Deutsch, C.V.: Projection pursuit multivariate transform. Math. Geosci. **46**, 337–360 (2014)
7. van den Boogaart, K. G., Tolosana-Delgado, R., Mueller, U. (2015). An affine equivariant anamorphosis for compositional data. In: Proceedings of IAMG 2015—17th Annual Conference of the International Association for Mathematical Geosciences, pp. 1302–1311 (2015)
8. Barnett, R.M., Deutsch, C.: Guide to multivariate modelling with the PPMT. Centre for Computational Geostatistics Guidebook Series, vol. 20 (2015)
9. Laparra, V., Camps-Valls, G., Malo, J.: Iterative Gaussianization: from ICA to random rotations. IEEE Trans. Neural Netw. Publ. IEEE Neural Netw. Council **22**, 537–549 (2011). https://doi.org/10.1109/TNN.2011.2106511
10. Enomoto, R., Okamoto, N., Seo, T.: Multivariate normality test using Srivastava's skewness and kurtosis. SUT J. Math. **48**(1), 103–115 (2012)
11. Tolosana-Delgado, R., Mueller, U., Gerald van den Boogart, K.: Geostatistis for compositional data: an overview. Math. Geosci. **51**(4), 485–526 (2019)
12. Szekely. G.J., Rizzo, M.L.: Energy statistics: a class of statistics based on distances. J. Stat. Plan. Infer. **143**(8), 1249–1272 (2013)

# Comparing and Detecting Stationarity and Dataset Shift

Camilla da Silva, Jed Nisenson, and Jeff Boisvert

**Abstract** Machine learning algorithms have been increasingly applied to spatial numerical modeling. However, it is important to understand when such methods will underperform. Machine learning algorithms are impacted by *dataset shift*; when modeling domains of interest present non-stationarities there is no guarantee that the trained models are effective in unsampled areas. This work aims to compare the stationarity requirement of geostatistical methods to the concept of dataset shift. Also, workflow is developed to detect dataset shift in spatial data prior to modeling, this involves applying a discriminative classifier and a two sample Kolmogorv-Smirnov test to model areas. And, when required a lazy learning modification of support vector regression is proposed to account for dataset shift. The benefits of the lazy learning algorithm are demonstrated on the well-known non-stationary Walker Lake dataset and improves root mean squared error up to 25% relative to standard SVR approach, in areas where dataset shift is present.

**Keywords** Dataset shift · Locally weighted learning · Stationarity

## 1 Introduction

Machine learning (ML) algorithms have gained space in the mineral resource's modeling process [4, 9, 12]. One motivational aspect of such increase in data-driven models use is the claim that ML methods generate non-stationary estimates and help solve the issue of non-stationary domains ([12, 14]), but care must be taken in making such claims. If a model is developed to predict block values from blast area A, can the model be used with confidence on blast area B? The answer to this question is not

---

A modified and extended version of this work has been submitted to the Geostats 2021 Special Issue in the journal Mathematical Geosciences.

---

C. da Silva (✉) · J. Boisvert
University of Alberta, Edmonton, AB T6G 1H9, Canada
e-mail: cdasilva@ualberta.ca

J. Nisenson
Teck Resources Limited, Burrard Street. 550, Vancouver V6C 0B3, Canada

© The Author(s) 2023
S. A. Avalos Sotomayor et al. (eds.), *Geostatistics Toronto 2021*, Springer Proceedings in Earth and Environmental Sciences, https://doi.org/10.1007/978-3-031-19845-8_3

obvious and highlights practical issues that stem from data-driven models. When ML is used, specifically supervised learning, the goal is to infer an underlying relationship between input variables and target variables [13] so that values at unsampled locations are predicted from the modeled relationship. Nonetheless, it is assumed that the multivariate distributions of the training stage and the test stage are identical. In geological settings non-stationarity leads to changes within modeling domains and can generate a phenomenon known as *dataset shift* in ML, which compromises model performance. Therefore, to obtain an accurate and representative model, the presence of dataset shift should be verified [1, 3, 5, 11] and, if present, accounted for. To that end, two algorithms are proposed: the first detects and maps the dataset shift present in geological settings,the second is proposed to handle dataset shift and provide accurate final predictions. The results demonstrate the sub-optimality of ML methods in non-stationary geological domains when dataset shift is not accounted for.

## 2   Materials and Methods

Herein, two algorithms are developed to detect and handle dataset shift in geospatial context. The proposed algorithm to detect dataset shift is based on the assumptions presented by Gözüaçik et al. [7]. It considers a variable of interest $Y$ in domain $A$, with a local neighborhood $W$ of fixed size. The samples contained in $W$ are compared to (1) the global data distribution and (2) local samples in an adjacent neighborhood, $K$. The algorithm is performed in two steps: first, the algorithm compares the data from two adjacent neighborhoods ($W$ and $K$). Samples within $W$ are classified as 0 and samples within $K$ as 1. The size of the neighborhoods can be defined as a fixed radius from an anchor point. Samples from both neighborhoods are merged to create a binary slack variable ($\zeta$). Classification of the merged data is attempted using logistic regression, where $Y$ is used to classify $\zeta$, the classifier is fit to predict the class (0 or 1) based on the sample values. The classifier's ability to distinguish between classes is measured with the area under the receiver operator characteristic curve (AUC). $AUC \approx 0.5$ indicates that the classifier is unable to separate the two classes, samples in the two neighborhoods are not shifted. $AUC \approx 1.0$ indicates that the classifier can separate the two classes, the data distributions do not overlap and are shifted. Intermediate $AUC$ values indicate the distributions partially overlap; typically, a threshold of $AUC > 0.7$ is used to determine if the distributions are shifted [7]. The second step of the algorithm considers a two sample Kolmogorv-Smirnov test (2 K-S test) [10] on samples in $W$ and $K$. The nonparametric 2 K-S test verifies if two samples come from the same distribution. The common way to report and interpret the 2 K-S test is through the P-value. A critical region is calculated such that the probability of wrongfully rejecting the hypothesis that the samples originate from the same distribution is not more than a predetermined threshold ($\alpha$). If the P-value is lower than the threshold ($\alpha$), the distinctions are significant, and the hypothesis is rejected. To detect shift relative to the global distribution, the rational is the same as with two neighborhoods,however, a random sampling of the global distribution is considered to obtain a representative subset with a similar number of samples as the local neighborhood to avoid affecting classifier performance due oversampling one class. For the 2 K-S test, the local neighborhood must contain enough samples to reliably estimate the distributions. Combining the results of the discriminative classifier and the 2 K-S test results in 3 possible scenarios:

$$
\begin{cases}
2, if\, AUC > \tau_1 \text{ and } p-value < \tau_2, \text{agreement and shift is likely} \\
1, if\, AUC > \tau_1 \text{ and } p-value > \tau_2 \text{ or } AUC < \tau_1 \text{ and } p-value < \tau_2, \text{disagreement and shift is possible} \\
0, if\, AUC < \tau_1 \text{ and } p-value > \tau_2, \text{agreement and shift is unlikely}
\end{cases}
$$

where $\tau_1$ is the threshold on AUC for the discriminative classifier, and $\tau_2$ is the P-value for the 2 K-S test.

The algorithm proposed to account for dataset shift is locally weighted support vector regression (LWSVR), aiming to adjust the training process to specific properties of sub-regions in the input space [2]. This is performed by assigning different importance to data most relevant to the location being predicted. Based on the principals of local models and weighting training data based on relevance, Ellatar et al. [6] proposed an algorithm in which the SVR risk function is modified to account for data relevance. In SVR traditional formulation $C$ is a fixed regularization parameter defined a priori by the user, however, generalization error changes if $C$ is modified according to a metric of relevance. The LWSVR risk function becomes:

$$
min \frac{1}{2}\|\omega\|^2 + C_i \sum_{n=1}^{N} \left(\xi_n + \xi_n^*\right), C_i = \Omega(\mathbf{d})C \tag{1}
$$

$\Omega(\mathbf{d})$ is the weight calculated for each local neighborhood according to function dependent on the Mahalanobis distance considered inside each search neighborhood and a smoothing factor that controls the generalization range.

## 3 Results and Discussion

The workflow is demonstrated on the Walker Lake sample set containing 470 samples from variable V [8]. First the database is inspected for dataset shift, then the standard SVR approach and the proposed LWSVR algorithm are applied to make spatial predictions and quantify improvement in the presence of dataset shift. Then, simple kriging (SK) and SK with locally varying means (SK with LVM) are considered to draw a parallel analysis to the ML approaches. The shift detection algorithm is applied to the Walker Lake sample set to detect regions where dataset shift occurs. The global shift analysis compares local windowed distributions to the global distribution while the local shift analysis compares local windowed distributions to other nearby local windowed distributions. To that end, consider a set of local neighborhoods, $W_i, i = 1, \ldots, n$, of fixed size of 60 m and $n$ samples,compare this distribution to either the global distribution or an adjacent subregion ($K$) with 30 m overlap. Choice of threshold to determine a binary, shift versus no shift, decision is dataset specific. In this case study, a sensitivity analysis leads to a threshold for the AUC of 0.8 and P-value of 0.05. The dataset shift detection algorithm demonstrated regions where the dataset shift occurs between local and global distributions. Such regions may benefit from trend modeling in case of a geostatistical approach, or a lazy learning algorithm in ML context. Given that dataset shift is detected, the LWSVR and SVR algorithms are applied and evaluated. The SVR model is optimized with a grid search considering $1e^{-5} < C < 100$; $1e^{-5} < \epsilon < 0.1$ and $1e^{-5} < \gamma < 10$. Optimal SVR parameters are: $C = 0.53$, $\gamma = 6.57$ and $\epsilon = 0.0017$. For LWSVR, the $C$ parameter is dependent on the closest data samples to the location being predicted. The number of samples retained for the LWSVR model is 5, while $\gamma$ and $\epsilon$ are held constant at 1.0 and 0.1, respectively. For both LWSVR and SVR the input variables for the predictions are the X and Y sample coordinates and the target variable is V.

The results show SVR obtained a smoother estimate than the LWSVR algorithm which reduced bias in the under sampled low valued regions (Table 1). Similarly, SK does not explicitly account for non-stationarity and produces a smoother estimate than SK with LVM. Bias in the predicted mean is evaluated using the exhaustive database, statistics are compared to the true statistics rather than the declustered data (Table 1); SK with LVM and LWSVR have lower bias in the mean as they better honor local features. As expected, SK with LVM is more variable than SK, similarly LWSVR is more variable than SVR (Table 1). Models' performance is evaluated with a 10-fold cross validation considering root mean squared error (RMSE) for each fold. Both local methods, LWSVR and SK with LVM, outperform their global counterparts. LWSVR results in a 25% RMSE improvement over SVR. It would be tempting to compare LWSVR to SK with LVM, but that comparison is inappropriate as SK honors data and provides a different modeling paradigm than ML algorithms. Because SK honors data, the impact of non-stationarity on geostatistical algorithms (3% improvement in RMSE) is much less than the impact of dataset shift on ML algorithms (25% improvement in RMSE) (Table 2).

**Table 1** Comparison of model statistics relative to the true exhaustive statistics

|  | True exhaustive | SK with LVM | SK | LWSVR | SVR |
|---|---|---|---|---|---|
| Mean | 278.0 | 272.6 | 293.4 | 290.9 | 300.6 |
| Std. Dev | 249.8 | 218.5 | 187.0 | 206.2 | 188.7 |
| CV | 0.89 | 0.80 | 0.64 | 0.71 | 0.63 |
| % Deviation relative to true mean |  | −1.94% | +5.53% | +4.64% | +7.91% |

**Table 2** Ten–fold cross validation

| Fold | SK with LVM | SK | LWSVR | SVR |
|---|---|---|---|---|
| Average RMSE | 185 | 190 | 208 | 275 |

The case study analysis reflects the nature of supervised ML algorithms, and objectively demonstrate the impact of dataset shift generated from non-stationarities in geospatial data. If the statistics change significantly between the training locations and where the modeling is deployed to obtain predictions, the relations previously learned are inefficient and lead to a final model that is not representative of the true geological phenomena. In this case, non-stationarities have to be explicitly (i.e. a trend model) or implicitly (i.e. local learning) accounted for; one such model, LWSVR, was proposed to consider this. The algorithm proposed to map dataset shift helps improve the modeling framework by identifying areas of interest where global algorithms are likely to underperform. However, some limitations persist. Applying the automated shift detection algorithm in sparse settings is sensitive to neighborhood search parameterization and the number of samples. The discriminative classifier is optimized for local neighborhoods, requiring that each labeled class have sufficient samples to form a training and test set. The choice of number of samples for LWSVR to generate a prediction is important and impacts conditional bias. The weighting function used in LWSVR also impacts performance. If sampling is dense, the penalty C applied on LWSVR is higher and can lead to overfit local models; while the search strategy and weight function can be easily modified, it may require tuning. Another aspect that must be considered is 3-dimensional data; the search strategy, anisotropy, and the implementation of the weight function should be modified accordingly. Finally, the impact of local optimization of $\gamma$ and $\epsilon$ should be considered, this study focused on local optimization of C.

# 4   Conclusions

Clear benefits of data-driven algorithms include reduced parameterization and fewer subjective modeling decisions; however, non-stationary spatial features often result in dataset shift within spatial modeling domains of interest. In this case, the impact of dataset shift on spatial modeling shows the importance of local learning. Practitioners must account for nonstationary spatial features of interest and understand how algorithms learning processes are affected by dataset shift and sparse sampling. Many algorithms do not have analogous lazy learning spatial implementations, as presented herein for SVR; it is the responsibility of the practitioner to understand the limitations of the chosen algorithm and investigate the appropriateness of associated lazy learning implementations.

# References

1. Baier, L., Hofmann, M., Kuhl, N., Mohr, M., Satzger, G.: Handling concept drift in regression problems—the error intersection approach. Comput. Sci. Math. (2020). https://doi.org/10.30844/wi_2020_c1-baier
2. Bottou, E., Vapnik, V.: Local learning algorithms. Neural Comput. **4** (1992). https://doi.org/10.1162/neco.1992.4.6.888.
3. Cejnek, M., Bukovsky, I.: Concept drift robust adaptive novelty detection for data streams. Neurocomputing **309** (2018). https://doi.org/10.1016/j.neucom.2018.04.069
4. Dai, F., Zhou, Q., Lv, Z., Wang, X., Liu, G.: Spatial prediction of soil organic matter content integrating artificial neural network and ordinary kriging in Tibetan Plateau. Ecol. Ind. **45** (2014). https://doi.org/10.1016/j.ecolind.2014.04.003.
5. Diethe, T., Borchert, T., Thereska, E., Balle, B., Lawrence, N.: Continual learning in practice. In: 32nd Conference on Neural Information Processing Systems (2018)
6. Ellatar, E.E., Goulermas, J., Wu, Q.H.: Electric load forecasting based on locally weighted support vector regression. IEEE Trans. Syst. Man Cybern. Part C (Appl. Rev.) **40**(4) (2010). https://doi.org/10.1109/TSMCC.2010.2040176
7. Gözüaçik, O., Bonalb, H., Büyükçakir, A., Can, F.: Unsupervised concept drift detection with discriminative classifier. In: International Conference on Information and Knowledge Management (CIKM '19) (2019). https://doi.org/10.1145/3357384.338144
8. Isaaks, E.H., Srivastava, M.R.: An Introduction to Applied Geostatistics. Oxford University Press, New York (1989)
9. Maniar, H., Ryali, R., Kulkarni, M.S., Abubakar, A.: Machine-learning methods in geoscience. In: SEG Technical Program Expanded Abstracts 2018 (2018). https://doi.org/10.1190/segam/2018-2997218.1
10. Pratt, J.W., Gibbons, J.D.: Concepts of nonparametric theory. Springer Series in Statistics. Springer, New York (1981)
11. Ruano-Ordás, D., Fdez-Riverola, F., Mendez, J.R.: Concept drift in e-mail datasets: an empirical study with practical implications. Inf. Sci. **428** (2018). https://doi.org/10.1016/j.ins.2017.10.049
12. Samson, M., Deutsch, C.V.: A hybrid estimation technique using elliptical radial basis neural networks and cokriging. Math. Geosci. (2021). https://doi.org/10.1007/s11004-021-09969-3
13. Sugiyama, M., Kawanabe, M.: Machine Learning in Non-Stationary Environments: Introduction to Covariate Shift Adaptation (Adapative Computation and Machine Learning Series). Massachussets institute of Technology, Cambridge (2012)
14. Shi, C., Wang, Y.: Non-parametric machine learning methods for interpolation of spatially varying non-stationary and non-Gaussian geotechnical properties. Geosci. Front. **12**(1), 339–350 (2021)

# Simulation of Stationary Gaussian Random Fields with a Gneiting Spatio-Temporal Covariance

Denis Allard, Xavier Emery, Céline Lacaux, and Christian Lantuéjoul

**Abstract** The nonseparable Gneiting covariance has become a standard to model spatio-temporal random fields. Its definition relies on a completely monotone function associated with the spatial structure and a conditionally negative semidefinite function associated with the temporal structure. This work addresses the problem of simulating stationary Gaussian random fields with a Gneiting-type covariance. Two algorithms, in which the simulated field is obtained through a combination of cosine waves are presented and illustrated with synthetic examples. In the first algorithm, the temporal frequency is defined on the basis of a temporal random field with stationary Gaussian increments, whereas in the second algorithm the temporal frequency is drawn from the spectral measure of the covariance conditioned to the spatial frequency. Both algorithms perfectly reproduce the correlation structure with minimal computational cost and memory footprint.

**Keywords** Substitution random fields · Spectral simulation · Spectral measure · Central limit approximation

## 1 Introduction

The modeling, prediction and simulation of stationary random fields defined on Euclidean spaces crossed with the time axis, $\mathbb{R}^k \times \mathbb{R}$ with, in general, $k = 2$ or 3, is widespread in hydrology, environment, climate, ecology and epidemiology appli-

D. Allard (✉)
INRAE (BioSP), 84914 Avignon, France
e-mail: denis.allard@inrae.fr

X. Emery
University of Chile, Santiago, Chile

C. Lacaux
Avignon University, LMA EA 2151, 84000 Avignon, France

C. Lantuéjoul
Mines Paris, PSL University, Fontainebleau, France

© The Author(s) 2023
S. A. Avalos Sotomayor et al. (eds.), *Geostatistics Toronto 2021*, Springer Proceedings in Earth and Environmental Sciences, https://doi.org/10.1007/978-3-031-19845-8_4

cations. The representation of their correlation structure via traditional covariance models in $\mathbb{R}^{k+1}$ is often unsuitable to capture space-time interactions, reason for which specific models need to be developed. One of these, the Gneiting covariance, is widely used in climate studies due to its versatility, and is defined as

$$C(\boldsymbol{h}, u) = \frac{\sigma^2}{\left(1 + \gamma(u)\right)^{k/2}} \, \varphi\left(\frac{|\boldsymbol{h}|^2}{1 + \gamma(u)}\right), \qquad \boldsymbol{h} \in \mathbb{R}^k, \, u \in \mathbb{R}, \qquad (1)$$

where $\sigma > 0$, $\gamma$ is a variogram (i.e., a conditionally negative semidefinite function) on $\mathbb{R}$ and $\varphi$ is a completely monotone function on $\mathbb{R}_+$. A subclass originally proposed by Gneiting in [1] is obtained by considering a variogram $\gamma$ of the form $\gamma(u) = \psi(u^2) - 1$, where $\psi$ is a Bernstein function, i.e., a positive primitive of a completely monotone function. The general formulation (1), in which $\gamma$ can be any variogram on $\mathbb{R}$, is due to Zastavnyi and Porcu in [2]. Hereinafter, without loss of generality, we assume $\sigma = 1$ and $\varphi(0) = 1$.

This work deals with the problem of simulating a stationary Gaussian random field with zero mean and Gneiting covariance on a (structured or unstructured) grid of $\mathbb{R}^k \times \mathbb{R}$. The following section presents some theoretical results, which will be used in Sects. 3 and 4 to design two simulation algorithms, which will be illustrated on synthetic examples. Concluding remarks follow in Sect. 5.

## 2   Theoretical Results

The completely monotone function $\varphi$ can be written as a nonnegative mixture of decreasing exponential functions on $\mathbb{R}_+$:

$$\varphi(t) = \varphi(0) \int_{\mathbb{R}_+} \exp(-r\,t) \, \mu(dr), \qquad t \in \mathbb{R}_+, \qquad (2)$$

where $\mu$ is a probability measure. Also, the continuous Fourier transform of a squared exponential function in $\mathbb{R}^k$ is another squared exponential function:

$$\int_{\mathbb{R}^k} \cos(\langle \boldsymbol{\omega}, \boldsymbol{h} \rangle) \, \exp(-a\,|\boldsymbol{\omega}|^2) d\boldsymbol{\omega} = \left(\frac{\pi}{a}\right)^{k/2} \exp\left(-\frac{|\boldsymbol{h}|^2}{4a}\right) \qquad \boldsymbol{h} \in \mathbb{R}^k, \, a > 0, \qquad (3)$$

where $\langle \cdot, \cdot \rangle$ stands for the usual scalar product in $\mathbb{R}^k$.

**Proposition 1** *By combining Eqs. (2) and (3), one can rewrite the Gneiting covariance (1) as follows:*

$$C(\boldsymbol{h}, u) = \int_{\mathbb{R}} \int_{\mathbb{R}^k} \cos(\sqrt{2r}\,\langle \boldsymbol{\omega}, \boldsymbol{h} \rangle) \, \exp\left(-\frac{\gamma(u)}{2}\,|\boldsymbol{\omega}|^2\right) g_{I_k}(\boldsymbol{\omega}) \, d\boldsymbol{\omega} \, \mu(dr), \qquad (4)$$

*with $I_k$ the identity matrix of order $k$ and $g_{I_k}$ the probability density of a $k$-dimensional Gaussian random vector with zero mean and covariance matrix $I_k$.*

The mappings $\boldsymbol{h} \mapsto \cos(\sqrt{2r}\,\langle \boldsymbol{\omega}, \boldsymbol{h}\rangle)$ and $u \mapsto \exp(-\gamma(u)\,|\boldsymbol{\omega}|^2/2)$ are covariances functions in $\mathbb{R}^k$ and $\mathbb{R}$, respectively. Their product is therefore a covariance function in $\mathbb{R}^k \times \mathbb{R}$, and so is $C(\boldsymbol{h}, u)$ as a positive mixture of covariances functions in $\mathbb{R}^k \times \mathbb{R}$. This result proves that every member of the Gneiting class (1) is a valid space-time covariance. In particular $C$ is a positive semidefinite function in $\mathbb{R}^k \times \mathbb{R}$.

**Proposition 2** *One can further decompose the Gneiting covariance as follows:*

$$C(\boldsymbol{h}, u) = \mathbb{E}\left\{\cos\left(\sqrt{2R}\,\langle \boldsymbol{\Omega}, \boldsymbol{h}\rangle + Y\,\sqrt{\gamma(u)}\,|\boldsymbol{\Omega}|\right)\right\}, \tag{5}$$

*where $\mathbb{E}\{\cdot\}$ the mathematical expectation, $R$ a nonnegative random variable with distribution $\mu$, $Y$ a standard normal random variable, $\boldsymbol{\Omega}$ a $k$-dimensional standardized Gaussian random vector, and where $R, Y, \boldsymbol{\Omega}$ are independent.*

**Proof** One uses (3) to write the squared exponential function in (4) as a Fourier transform on $\mathbb{R}$:

$$C(\boldsymbol{h}, u) = \int_{\mathbb{R}} \int_{\mathbb{R}^k} \int_{\mathbb{R}} \cos(\sqrt{2r}\,\langle \boldsymbol{\omega}, \boldsymbol{h}\rangle)\,\cos(y\sqrt{\gamma(u)}\,|\boldsymbol{\omega}|)\,g_1(y)\,g_{I_k}(\boldsymbol{\omega})\,dy\,d\boldsymbol{\omega}\,\mu(dr),$$

where $g_1$ is the standard normal univariate probability density function. Owing to the parity of this function and to the product-to-sum trigonometric identity, this expression simplifies into

$$C(\boldsymbol{h}, u) = \int_{\mathbb{R}} \int_{\mathbb{R}^k} \int_{\mathbb{R}} \cos(\sqrt{2r}\,\langle \boldsymbol{\omega}, \boldsymbol{h}\rangle + y\sqrt{\gamma(u)}\,|\boldsymbol{\omega}|)\,g_1(y)\,g_{I_k}(\boldsymbol{\omega})\,dy\,d\boldsymbol{\omega}\,\mu(dr),$$

which yields the claim.

# 3 A Discrete-in-Time and Continuous-in-Space Substitution Algorithm

Consider a space-time cosine wave of the following form:

$$Z(\boldsymbol{x}, t) = \sqrt{2} \cos\left(\sqrt{2R}\,\langle \boldsymbol{\Omega}, \boldsymbol{x}\rangle + W(t)\,\frac{|\boldsymbol{\Omega}|}{\sqrt{2}} + \Phi\right), \qquad \boldsymbol{x} \in \mathbb{R}^k, t \in \mathbb{R}, \tag{6}$$

where

- $R$ and $\boldsymbol{\Omega}$ are a random variable and a random vector as defined in (5);
- $\{W(t) : t \in \mathbb{R}\}$ is a strictly intrinsic random field with variogram $\gamma$ and Gaussian increments;

- $\Phi$ is a uniform random variable on $]0, 2\pi[$;
- $R$, $\mathbf{\Omega}$, $W$ and $\Phi$ are independent.

Because $\Phi$ is uniform on $]0, 2\pi[$ and is independent of $(R, \mathbf{\Omega}, W)$, this cosine wave is centered. Moreover, the covariance between $Z(\mathbf{x} + \mathbf{h}, t + u)$ and $Z(\mathbf{x}, t)$ is found to be equal to the expectation in (5), that is, $C(\mathbf{h}, u)$ (see [3]). The random field in (6) is a particular case of substitution random field, consisting of the composition of a stationary coding process on $\mathbb{R}$ and an intrinsic directing function on $\mathbb{R}^k \times \mathbb{R}$ (see [4]).

To obtain an approximately Gaussian random field with zero mean and Gneiting covariance, one can (i) multiply the cosine by a Rayleigh random variable with scale parameter $2^{-1/2}$, which makes the marginal distribution of $Z(\mathbf{x}, t)$ be standard Gaussian, and (ii) sum and standardize many of such independent cosine waves, so that the finite-dimensional distributions of $Z(\mathbf{x}, t)$ become approximately multivariate Gaussian due to the central limit theorem. The simulated random field thus takes the form:

$$
Z(\mathbf{x}, t) = \sum_{j=1}^{p} \sqrt{\frac{-2 \ln U_j}{p}} \cos \left( \sqrt{2R_j} \langle \mathbf{\Omega}_j, \mathbf{x} \rangle + W_j(t) \frac{|\mathbf{\Omega}_j|}{\sqrt{2}} + \Phi_j \right), \quad (7)
$$

where $p$ is a large integer, $\{(R_j, \mathbf{\Omega}_j, W_j, \Phi_j) : j = 1, ..., p\}$ are independent copies of $(R, \mathbf{\Omega}, W, \Phi)$, and $\{U_j : j = 1, ..., p\}$ are independent random variables uniformly distributed on $]0, 1[$ and are independent of $\{(R_j, \mathbf{\Omega}_j, W_j, \Phi_j) : j = 1, ..., p\}$.

As an illustration, consider the simulation of a random field on a regular grid of $\mathbb{R}^1 \times \mathbb{R}$ with $500 \times 500$ nodes and mesh $1 \times 0.2$, with the following parameters:

- $k = 1$;
- $\varphi(r) = \exp(-0.001r)$
- $\gamma(u) = \sqrt{1 + |u|} - 1$;
- $p = 10$, $100$ or $1000$.

The intrinsic random field $W$ is simulated by using the covariance matrix decomposition algorithm with the nonstationary covariance function $(t, t') \mapsto \gamma(t) + \gamma(t') - \gamma(t' - t)$. The simulation obtained with $p = 10$ cosine waves exhibits an apparent periodicity in space, which indicates that the central limit approximation is poor, which is no longer the case when using 100 or more cosine waves (Fig. 1, left). In contrast, since the simulated process $W$ is ergodic, the time variations are well reproduced, irrespective of the number of cosine waves. Note that the simulated random field has a Gaussian spatial covariance and a gamma temporal covariance with parameter 0.5, hence it is smooth in space but not in time.

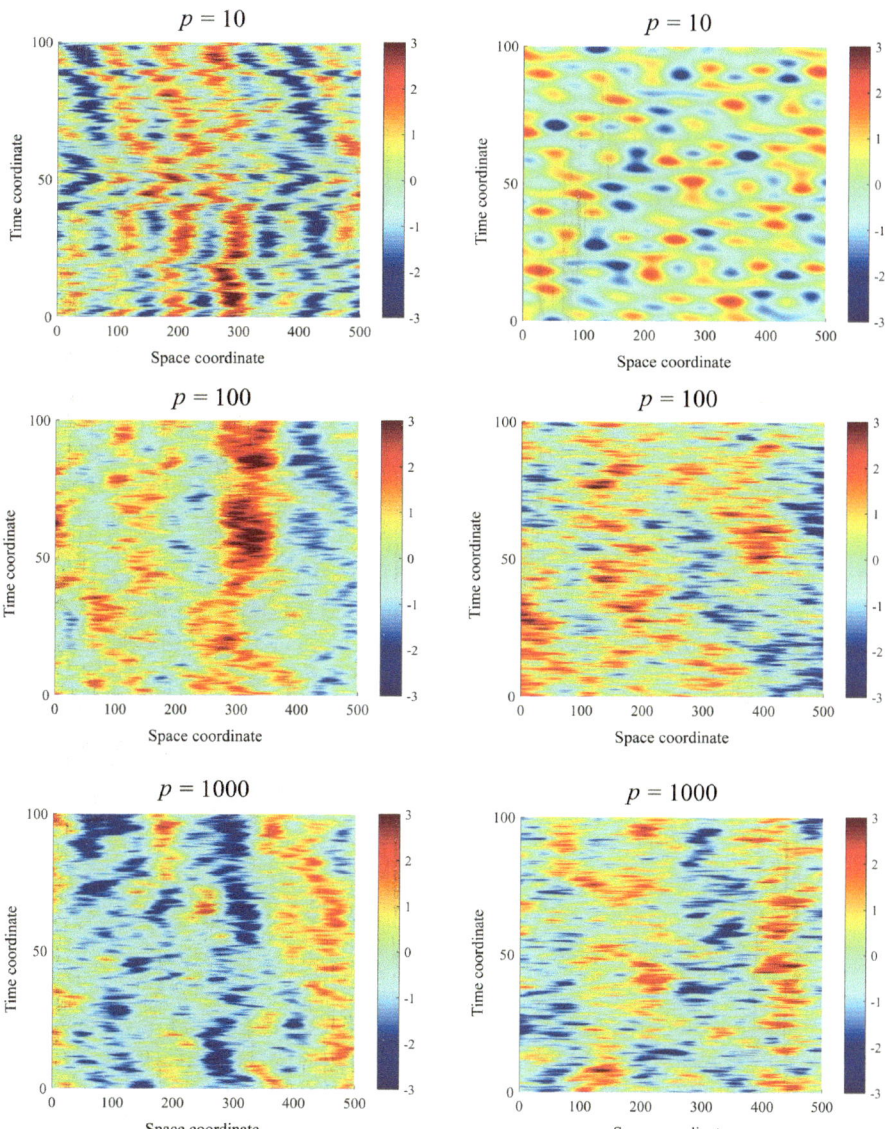

**Fig. 1** Simulation of a random field with Gneiting covariance function (with an exponential function for $\varphi$ and a square root for $\gamma$) using 10 (top), 100 (center) and 1000 (bottom) cosine waves. Left: substitution algorithm. Right: spectral algorithm

## 4   A Fully Continuous Spectral Algorithm

The continuous spectral method relies on the fact that the continuous covariance $C$ is the Fourier transform of a symmetric probability measure $F$ on $\mathbb{R}^k \times \mathbb{R}$. Let $(\boldsymbol{\Omega}, \mathrm{T})$ be a spectral vector distributed according to $F$, and $\Phi$ a random phase uniform on $]0, 2\pi[$ and independent of $(\boldsymbol{\Omega}, \mathrm{T})$. Then, the random field defined by

$$Z(\boldsymbol{x}, t) = \sqrt{2}\cos(\langle \boldsymbol{\Omega}, \boldsymbol{x}\rangle + \mathrm{T}t + \Phi), \qquad (\boldsymbol{x}, t) \in \mathbb{R}^k \times \mathbb{R}, \tag{8}$$

is second-order stationary with covariance $C$. A standard approach for simulating the spectral vector is to simulate at first the spatial component $\boldsymbol{\Omega}$, then the temporal component T given $\boldsymbol{\Omega}$, which requires explicitly knowing the spectral measure. For instance, consider the covariance function with $\varphi(r) = \exp(-ar)$ $(a > 0)$ and $\gamma(u) = \sqrt{1 + |u|} - 1$. In this case, it can be shown (see [3]) that $\boldsymbol{\Omega}$ is a Gaussian random vector with independent components and that T given $\boldsymbol{\Omega} = \omega$ follows a Cauchy distribution whose scale parameter follows an inverse Gaussian distribution, all these distributions being simulatable.

In practice, the multiplication of the cosine wave (8) by a Rayleigh random variable and the independent replication technique of (7) provide a random field with Gaussian marginal distributions and approximately multivariate Gaussian finite-dimensional distributions. Figure 1 (right) displays realizations obtained by using between $p = 10$ and $p = 1000$ cosine waves. The spatial variations are similar to those observed with the substitution algorithm. However, the temporal variations differ when using few cosine waves, exhibiting a smooth and periodic behavior when $p = 10$ or $p = 100$. This suggests that the convergence to a multivariate-Gaussian distribution in time is slower with the spectral algorithm than with the substitution algorithm.

## 5   Concluding Remarks

The two presented approaches construct the simulated random field as a weighted sum of cosine waves with random frequencies and phases. Their main difference lies in the way to simulate the temporal frequency: from its distribution conditional to the spatial frequency (spectral approach), or from an intrinsic time-dependent random field (substitution approach). Both algorithms have a computational complexity in $\mathcal{O}(n)$, considerably cheaper than generic algorithms such as the covariance matrix decomposition and sequential algorithms, are parallelizable and require minimal memory storage space, which makes them affordable for large-scale problems. The substitution approach is general and only requires the knowledge of the temporal variogram $\gamma$ and the probability measure $\mu$ specifying the completely monotone function $\varphi$ (2) associated with the spatial structure. In contrast, the spectral approach is more specific, as it also requires the knowledge of the time frequency distribution conditional to the spatial frequency, which has to be solved on a case-by-case basis.

**Acknowledgements** Denis Allard and Christian Lantuéjoul acknowledge support of the RESSTE network (https://reseau-resste.mathnum.inrae.fr) funded by the MathNum division of INRAE. Xavier Emery acknowledges the support of grants ANID PIA AFB180004 and ANID FONDE-CYT REGULAR 1210050 from the National Agency for Research and Development of Chile.

# References

1. Gneiting, T.: Nonseparable, stationary covariance functions for space-time data. J. Am. Stat. Assoc. **97**(458), 590–600 (2002)
2. Zastavnyi, V.P., Porcu, E.: Characterization theorems for the Gneiting class of space-time covariances. Bernoulli **17**(1), 456–465 (2011)
3. Allard, D., Emery, X., Lacaux, C., Lantuéjoul, C.: Simulating space-time random fields with nonseparable Gneiting-type covariance functions. Stat. Comput. **30**(5), 1479–1495 (2020)
4. Lantuéjoul, C.: Geostatistical Simulation: Models and Algorithms. Springer, Berlin (2002)

# Spectral Simulation of Gaussian Vector Random Fields on the Sphere

Alfredo Alegría, Xavier Emery, Xavier Freulon, Christian Lantuéjoul, Emilio Porcu, and Didier Renard

**Abstract** Isotropic Gaussian random fields on the sphere are used in astronomy, geophysics, oceanography, climatology and remote sensing applications. However, to date, there is a lack of simulation algorithms that reproduce the spatial covariance structure without any approximation and, at the same time, are parsimonious in terms of computation time and memory storage requirements. This work presents two such algorithms that rely on the spectral representation of isotropic covariances on the sphere. Both algorithms are illustrated with synthetic examples.

**Keywords** Isotropic random fields · Spherical harmonics · Legendre polynomials · Schoenberg sequence

## 1 Introduction

Random fields defined on the unit sphere $\mathbb{S}^2 = \{s \in \mathbb{R}^3 : |s| = 1\}$ are used in astronomy, geophysics, geotechnics, oceanography, climatology and remote sensing applications, where it is frequent to deal with multivariate data. Under an assumption of isotropy and multivariate normality, the only parameters to infer are the first-order moment (expectation vector), constant over the sphere and hereafter assumed to be zero, and the second-order moment (scalar or matrix-valued covariance function), which only depends on the geodesic distance $\delta$ between any two points on the sphere.

A. Alegría (✉)
Universidad Técnica Federico Santa María, Valparaíso, Chile
e-mail: alfredo.alegria@usm.cl

X. Emery
University of Chile, Santiago, Chile

X. Freulon · C. Lantuéjoul · D. Renard
Mines Paris, PSL University, Fontainebleau, France

E. Porcu
Khalifa University, Abu Dhabi, United Arab Emirates

© The Author(s) 2023

S. A. Avalos Sotomayor et al. (eds.), *Geostatistics Toronto 2021*, Springer Proceedings in Earth and Environmental Sciences, https://doi.org/10.1007/978-3-031-19845-8_5

Regrettably, although many computationally efficient algorithms are available to accurately simulate Gaussian random fields in Euclidean spaces, the same does not occur with random fields defined on the sphere. Simulation achieved through approximations into cosine waves or into spherical harmonics often reproduce the spatial correlation structure approximately. The objective of this work is to present two algorithms to simulate isotropic Gaussian random fields on $\mathbb{S}^2$ that exactly reproduce the target covariance and are efficient from a computational standpoint. Both algorithms rely on the spectral representation of isotropic covariances on the sphere, which is reminded in the next section.

## 2 Mathematical Background

According to Yaglom in [1], the covariance function of an isotropic vector random field on the sphere can be expanded as follows:

$$C\big(\delta(s, s')\big) = \sum_{k=0}^{+\infty} B_k \, P_k(s \cdot s'), \qquad s, s' \in \mathbb{S}^2, \tag{1}$$

where $\cdot$ is the usual scalar product in $\mathbb{R}^3$, $P_k$ is the Legendre polynomial of degree $k$ and $(B_k : k \in \mathbb{N})$ is a sequence of real-valued, symmetric, positive semidefinite matrices, called Schoenberg matrices, that are componentwise summable, i.e., such that $C(0) = \sum_{k=0}^{+\infty} B_k$ exists.

For any $s \in \mathbb{S}^2$ with colatitude $\theta \in [0, \pi]$ and longitude $\phi \in [0, 2\pi]$, the spherical harmonics of degree $k \in \mathbb{N}$ and order $m \in \{-k, \ldots, k\}$ is defined as:

$$Y_{k,m}(s) = (-1)^m \sqrt{\frac{(2k+1)(k-)}{4\pi(k+)}} \, P_k(\cos\theta) \times \begin{cases} \sqrt{2}\sin(\phi) & \text{if } m < 0 \\ 1 & \text{if } m = 0 \\ \sqrt{2}\cos(\phi) & \text{if } m > 0 \end{cases} \tag{2}$$

where $P_k^m$ is the associated Legendre function of degree $k$ and order $m$. The spherical harmonics satisfy the following two properties.

(1) Addition theorem:

$$\frac{4\pi}{2k+1} \sum_{m=-k}^{+k} Y_{k,m}(s) Y_{k,m}(s') = P_k(s \cdot s'), \qquad k \in \mathbb{N}. \tag{3}$$

(2) Orthogonality:

$$4\pi \int_{\mathbb{S}^2} Y_{k,m}(s) Y_{k',m'}(s) \, U(ds) = \begin{cases} 1 & \text{if } k = k' \text{ and } m = m' \\ 0 & \text{otherwise,} \end{cases} \tag{4}$$

where $k, k' \in \mathbb{N}, m \in \{-k, \ldots, +k\}, m' \in \{-k', \ldots, +k'\}$, and $U$ is the uniform distribution on $\mathbb{S}^2$.

# 3  Simulation Algorithms

## 3.1  Random Mixture of Spherical Harmonics (RMSH)

Let $f$ be a probability mass function on $\mathbb{N}$ such that $f(k) > 0$ whenever $B_k$ is not a zero matrix. If $K \sim f$, then Schoenberg's formula (1) becomes

$$C\big(\delta(s, s')\big) = \mathbb{E}\Big\{\frac{B_K}{f(K)}\, P_K(s \cdot s')\Big\}. \tag{5}$$

Moreover, if $M$ is uniform over $\{-K, \ldots, +K\}$, then the addition formula (3) gives

$$P_K(s \cdot s') = 4\pi\, \mathbb{E}\big\{Y_{K,M}(s)\, Y_{K,M}(s') \,\big|\, K\big\}. \tag{6}$$

Then, combining Eqs. (5) and (6), one obtains

$$C\big(\delta(s, s')\big) = 4\pi\, \mathbb{E}\Big\{\frac{B_K}{f(K)}\, Y_{K,M}(s)\, Y_{K,M}(s')\Big\}.$$

Letting $A_k$ be a symmetric square root of $B_k$ ($k \in \mathbb{N}$) and $A_k(\cdot, J)$ be the $J$-th column of $A_k$, with $J$ an integer uniform over $\{1, \ldots, p\}$, one furthermore has:

$$\mathbb{E}\Big\{A_k(\cdot, J)A_k(\cdot, J)^\top\Big\} = \frac{1}{p}\sum_{j=1}^{p} A_k(\cdot, j)A_k(\cdot, j)^\top = \frac{1}{p}B_k.$$

The previous equations suggest the following construction for simulating a random field on $\mathbb{S}^2$ with covariance $C$:

$$\tilde{Z}(s) = \varepsilon\sqrt{\frac{4\pi\, p}{f(K)}}\, A_K(\cdot, J)\, Y_{K,M}(s), \qquad s \in \mathbb{S}^2, \tag{7}$$

with $\varepsilon$ a random variable with zero mean and unit variance independent of $(K, M)$ and $J$ an integer uniform over $\{1, \ldots, p\}$ and independent of $(\varepsilon, K, M)$.

## 3.2 Random Mixture of Legendre Waves (RMLW)

The second simulation algorithm rests on the following identity:

$$\int_{\mathbb{S}^2} P_k(\omega \cdot s) \, P_k(\omega \cdot s') \, U(d\omega) = \frac{1}{2k+1} \, P_k(s \cdot s'), \qquad k \in \mathbb{N}, \tag{8}$$

which can be derived from the addition theorem (3) and the orthogonality of spherical harmonics (4). Equation (8) can be rewritten in probabilistic terms:

$$\frac{1}{2k+1} \, P_k(s \cdot s') = \mathbb{E}\big\{ P_k(\Omega \cdot s) \, P_k(\Omega \cdot s') \big\}, \qquad k \in \mathbb{N}, \tag{9}$$

where $\Omega$ is a random point (pole) uniformly distributed on $\mathbb{S}^2$. The covariance function (5) of an isotropic vector random field $Z$ becomes:

$$C\big(\delta(s, s')\big) = \mathbb{E}\left\{ \frac{2K+1}{f(K)} \, B_K \, P_K(\Omega \cdot s) P_K(\Omega \cdot s') \right\}, \qquad s, s' \in \mathbb{S}^2,$$

where $K$ is a random integer with probability mass function $f$, independent of $\Omega$. Following the same reasoning as in the previous section, a random field $\tilde{Z}$ sharing the same first two moments as $Z$ is obtained by putting

$$\tilde{Z}(s) = \varepsilon \sqrt{\frac{(2K+1)p}{f(K)}} \, A_K(\cdot, J) \, P_K(\Omega \cdot s), \qquad s \in \mathbb{S}^2, \tag{10}$$

with $\varepsilon$ a random variable with zero mean and unit variance independent of $(K, \Omega)$, $J$ an integer uniform in $\{1, \ldots, p\}$ and independent of $(\varepsilon, K, \Omega)$, and $A_K(\cdot, J)$ the $J$-th column of a symmetric square root $A_K$ of the Schoenberg matrix $B_K$. The construction (10) has been named "turning arcs" by Alegría et al. in [2], as it is the exact analogue of the turning bands method in which a random field defined along a straight line is spread to the multidimensional Euclidean space; here, a Legendre wave $P_K(\Omega \cdot s)$ that is constant over the arcs perpendicular to $\Omega$ is spread to the sphere.

## 3.3 Discussion

The two previous proposals can be classified as continuous spectral algorithms, in which the simulated field is a basic random field (harmonic) defined continuously on the sphere, consisting of a spherical harmonic with random degree and order or a Legendre wave with random degree and pole. This basic harmonic is weighted by a random vector that ensures the reproduction of the target spatial correlation structure, which reminds of importance sampling techniques.

Both algorithms provide continuous representations of isotropic random fields on $\mathbb{S}^2$ with finite-dimensional distributions that are not multivariate Gaussian. A central limit approximation can be used to obtain a Gaussian random field, based on $L$ independent copies of $\tilde{Z}$ defined either by (7) or (10). The computational complexity is proportional to the number $L$ and the number $n$ of locations targeted for simulation, i.e., $\mathcal{O}(n \times L)$; this compares favorably with the covariance matrix decomposition algorithm, whose numerical complexity is proportional to $n^3$. Interestingly, both algorithms can be adapted to the simulation of isotropic random fields on the $d$-sphere, with $d > 2$, by replacing the spherical harmonics by hyperspherical harmonics in the RMSH algorithm, or the Legendre polynomial by a Gegenbauer polynomial in the RMLW algorithm. The validity of these adapted algorithms stems from the addition theorem and the orthogonality of hyperspherical harmonics.

## 4   Examples

As a first example, consider the univariate multiquadric covariance on the sphere:

$$C\big(\delta(s, s')\big) = \frac{1 - \mu}{\sqrt{1 - 2\mu \cos \delta(s, s') + \mu^2}}, \qquad s, s' \in \mathbb{S}^2,$$

whose Schoenberg sequence is the geometric probability mass function $(1 - \mu)\mu^k$ (see [3]). In the following we set $\mu = 0.7$ and discretize the sphere into $500 \times 500$ points with regularly-spaced colatitudes and longitudes. Both algorithms are applied to generate one realization using $L = 10$ and $100$ basic random fields, with $\varepsilon$ following a Rademacher distribution and $K + 1$ having a zeta distribution with parameter 2. The latter distribution is long tailed and allows sampling high degree harmonics ($K$ large) with a non-negligible probability. The realizations obtained by both algorithms look the same when the number of basic random fields is high ($L \geq 100$), which suggests that the central limit approximation is acceptable for such a number of basic random fields (Fig. 1).

The second example is the univariate Chentsov covariance:

$$C\big(\delta(s, s')\big) = 1 - \frac{2\delta(s, s')}{\pi}, \qquad s, s' \in \mathbb{S}^2.$$

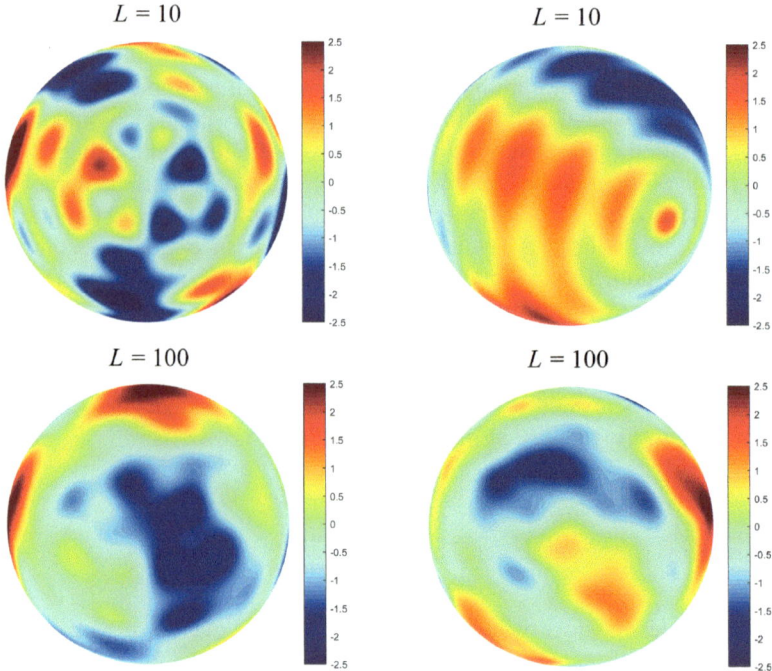

**Fig. 1** Realizations of a scalar random field with a multiquadric covariance (parameter $\mu = 0.7$), constructed with the RMSH (left) and RMLW (right) algorithms, for $L = 10$ and 100 basic random fields.

The associated Schoenberg sequence is (see [4]):

$$b_k = \begin{cases} 0 & \text{if } k \text{ is even} \\ \frac{2k+1}{4\pi} \frac{\Gamma^2(k/2)}{\Gamma^2((k+3)/2)} & \text{if } k \text{ is odd.} \end{cases}$$

Again, the two algorithms are applied to generate one realization using $L = 1000$ and 10,000 basic random fields, and considering the same discretization of the sphere and the same distributions for $K$ and $\varepsilon$ (Fig. 2). The convergence to normality turns out to be slower here, which is explained because the Chentsov covariance corresponds to a random field that is continuous but not differentiable, whereas the spherical harmonics and Legendre waves are smooth functions: many such functions ($L \geq$ 10,000) are necessary to sample the tail of the zeta distribution sufficiently to reproduce the short-scale behavior of the target random field. With fewer functions, a striation effect is perceptible in the realizations.

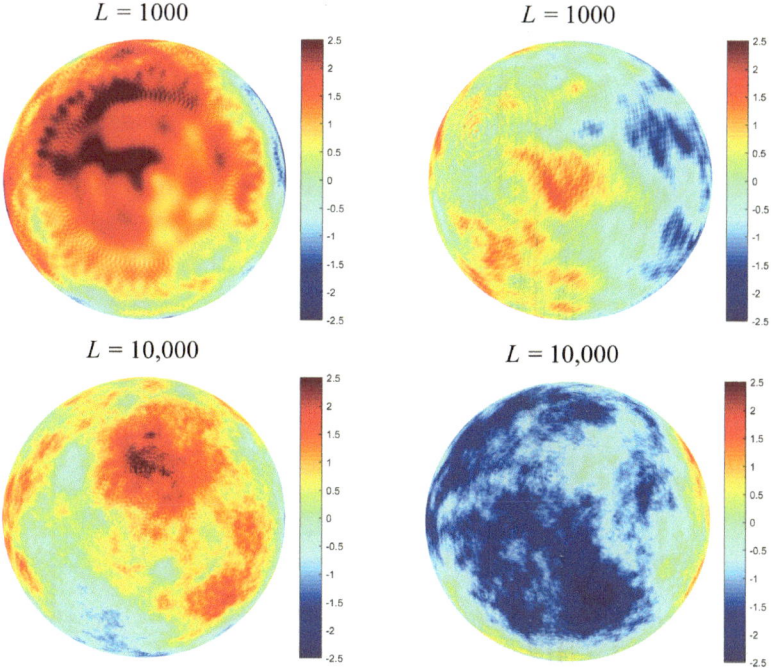

**Fig. 2** Realizations of a scalar random field with a Chentsov covariance, constructed with the RMSH (left) and RMLW (right) algorithms, for $L = 1000$ and $10,000$ basic random fields.

The last example is a bivariate ($p = 2$) spectral Matérn covariance, defined through its Schoenberg matrices (see [3]):

$$B_k = \begin{bmatrix} S(v_{11})^{-1}(1 + k^2)^{-v_{11}-1/2} & \rho S(v_{12})^{-1}(1 + k^2)^{-v_{12}-1/2} \\ \rho S(v_{12})^{-1}(1 + k^2)^{-v_{12}-1/2} & S(v_{22})^{-1}(1 + k^2)^{-v_{22}-1/2} \end{bmatrix}, \qquad k \in \mathbb{N},$$

with $v_{11} > 0$, $v_{22} > 0$, $v_{12} = \frac{v_{11}+v_{22}}{2}$, $|\rho| \leq 1$ and $S(v) = \sum_{k=0}^{+\infty}(1 + k^2)^{-v-1/2}$.

We set $\rho = -0.9$, $v_{11} = 0.75 < 1$ and $v_{22} = 1.25 > 1$, so that the second random field component is mean-square differentiable, while the first component is not. Figure 3 shows one realization obtained with the RMLW algorithm by using a zeta distribution for $K + 1$ and a Rademacher distribution for $\varepsilon$, for $L = 10,000$ (similar results are obtained with the RMSH algorithm and are not displayed here). As expected, the first component is irregular whereas the second is smooth, both components being negatively correlated ($\rho = -0.9$). The striation effect is hardly perceptible, suggesting that the chosen number of basic random fields is sufficient for the central limit approximation to be acceptable.

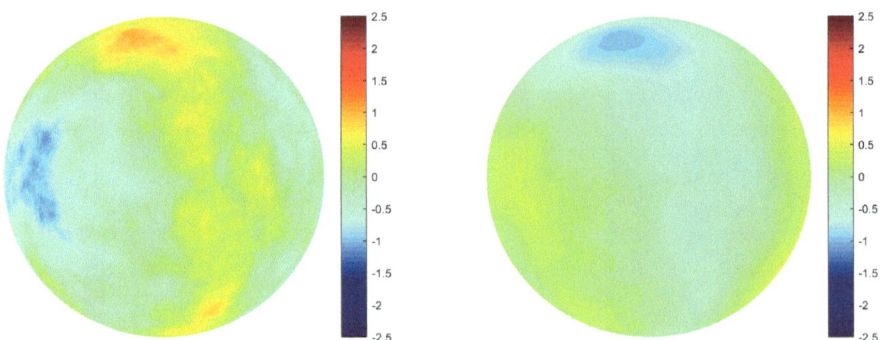

**Fig. 3** One realization of a bivariate random field with a spectral Matérn covariance (parameters $v_{11} = 0.75$ and $v_{22} = 1.25$), constructed with the RMLW algorithm and $L = 10{,}000$ basic random fields.

## 5 Conclusions

Two algorithms have been proposed to simulate vector Gaussian random fields on the two-dimensional sphere. Both rest on the spectral decomposition of the covariance function. They provide continuous simulations, in the sense that they start by building basic ingredients that subsequently allow computing the value of the simulated field at any point on the sphere. Moreover, they can be generalized to perform simulations on hyperspheres. Convergence to multivariate normality is reached with fewer basic random fields when using the RMSH algorithm in comparison with the RMLW algorithm, because spherical harmonics are comparatively more multichromatic than Legendre waves. Compensatorily, it takes less time to compute Legendre polynomials than spherical harmonics.

**Acknowledgements** The authors acknowledge the support of the National Agency for Research and Development of Chile, through Grants ANID PIA AFB180004 and ANID FONDECYT REGULAR 1210050.

## References

1. Yaglom, A.: Correlation Theory of Stationary and Related Random Functions: Basic Results. Springer, New York (1987). https://doi.org/10.1007/978-1-4612-4628-2
2. Alegría, A., Emery, X., Lantuéjoul, C.: The turning arcs: a computationally efficient algorithm to simulate isotropic vector-valued Gaussian random fields on the $d$-sphere. Stat. Comput. **30**(5), 1403–1418 (2020). https://doi.org/10.1007/s11222-020-09952-8
3. Emery, X., Porcu, E.: Simulating isotropic vector-valued Gaussian random fields on the sphere through finite harmonics approximations. Stoch. Environ. Res. Risk Assess. **33**(8–9), 1659–1667 (2019). https://doi.org/10.1007/s00477-019-01717-8
4. Lantuéjoul, C., Freulon, X., Renard, D.: Spectral simulation of isotropic Gaussian random fields on a sphere. Math. Geosci. **51**(8), 999–1020 (2019). https://doi.org/10.1007/s11004-019-09799-4

# Petroleum

# Geometric and Geostatistical Modeling of Point Bars

Ismael Dawuda and Sanjay Srinivasan

**Abstract** Point bar reservoir geology is frequently encountered in oil and gas developments worldwide. Furthermore, point bar geology is encountered in many sites being considered for large scale $CO_2$ injection for sequestration. A comprehensive modeling method that adequately preserves point bar internal architecture and its associated heterogeneities is still not available. Traditional geostatistical methods cannot adequately capture the curvilinear architecture of point bars. Even geostatistical simulation techniques that can be constrained to multiple point statistics cannot capture the architecture of the point bars because they use regular grids to represent the heterogeneity. If heterogeneities like the thinly distributed shale drapes within the point bar are represented using an extremely fine mesh, the computational cost for performing flow modeling escalates steeply. This paper proposes a modeling method that preserves the point bar internal architecture and heterogeneities, without these limitations. The modeling method incorporates a gridding scheme that adequately captures the point bar architecture and heterogeneities, without huge computational costs.

**Keywords** Simulation · Reservoir modeling · $CO_2$ sequestration · Local anisotropy

## 1 Introduction

Point bars (Fig. 1) are fluvial sediments that accumulate at the inner bends of channel meanders by deposition of eroded sediments as the channel migrates outwards [1, 20, 37]. Point bars have great economic significance [7], as they can serve as large storage reservoirs [14, 24]. The McMurray Formation in Alberta, Canada, which hosts large accumulations of bitumen is predominantly composed of point bar deposits [36]. Other examples are those of the Widuri Field [6] and the Little Creek Field in southwestern Mississippi [35].

I. Dawuda (✉) · S. Srinivasan
Energy and Mineral Engineering Department, The Pennsylvania State University, State College, USA
e-mail: iud30@psu.edu

© The Author(s) 2023 63
S. A. Avalos Sotomayor et al. (eds.), *Geostatistics Toronto 2021*, Springer Proceedings in Earth and Environmental Sciences, https://doi.org/10.1007/978-3-031-19845-8_6

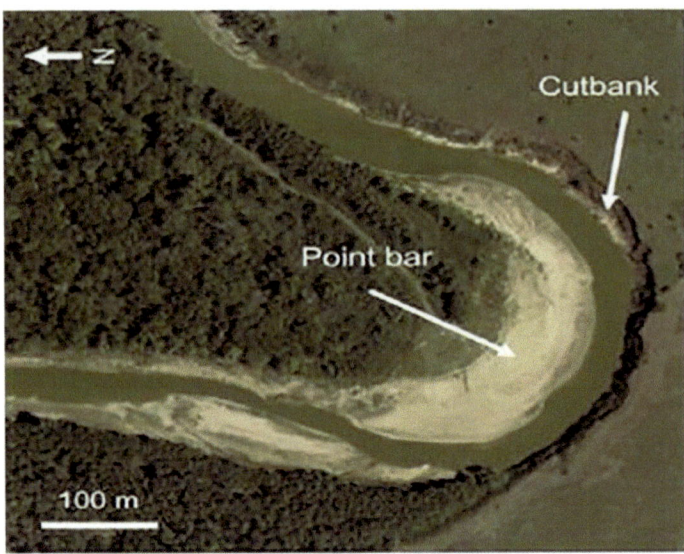

**Fig. 1** Point bar deposit, formed by erosion of sediments from the outer bend (cutbank), and deposition of the eroded sediments at the inner bend, adapted from [15]

However, point bars exhibit complex spatial distribution of heterogeneities [12, 24, 33]. As an example, [34] identified different forms of depositional trends in different directions within point bar deposits. Some of these directional trends include fining upwards, fining along downstream direction and fining in the direction perpendicular to the inclined layers (called inclined heterolithic stratification (IHS)). These trends can affect the exploitation of the subsurface for hydrocarbon production and geological storage of $CO_2$. For example, point bar heterogeneities like shale drapes along the IHS surfaces act as flow barriers, compartmentalize the reservoir and decrease storage capacity [16, 17]. Therefore, developing modeling methods for representing point bars is of economic consequence.

Several studies have used different methods to develop point bar models. Some of these methods are process-based (e.g., [29, 32], object-based (e.g., [4, 11, 39], surface-based (e.g., [26, 30], and geostatistical simulation methods like sequential indicator simulation (e.g., [9]. The geostatistical-based methods remain popular among researchers and modelers. More recently, [8] combined geometric modeling with geostatistical computations to represent point bar geometries and their petrophysical property distribution. This included the use of sine generation function (SGF) to model the aerial dimension of the point bar, to capture the lateral accretions. The main drawback of their study is that the use of the SGF may not yield realistic approximations for point bars with asymmetric geometries.

In this study, a cubic spline function is used to develop a smooth geometric model of the point bar that captures the lateral accretions, while a sigmoidal function is used to model the inclined heterolithic stratifications (IHS). A key element of

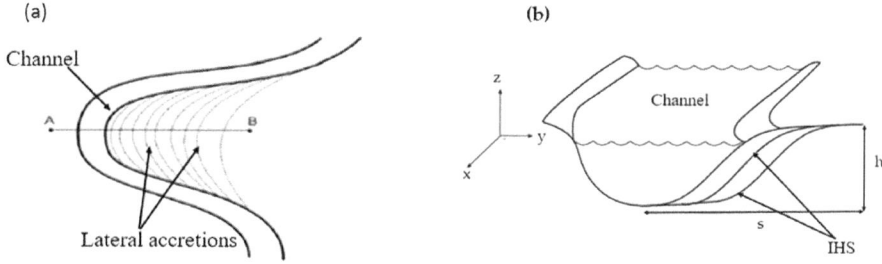

**Fig. 2** Point bar schematic, showing **a** the lateral accretions and **b** the inclined heterolithic stratifications (IHS). Adapted from [31, 34]

the modeling approach is the incorporation of a computationally inexpensive grid generation scheme that preserves the point bar curvilinear architecture.

## 2  An Overview of Point Bar Geometry

The main heterogeneities in the point bars are the lateral accretions and the inclined heterolithic stratifications (IHS). These heterogeneities are formed by episodic migration of meandering channels, due to the erosion of sediments from the outer bend of the channel, and deposition of the eroded sediments into the inner bend. If one moves along section AB in Fig. 2a, a channel is first encountered. A further progression towards point B shows some curvy structures. Those are the lateral accretions, which are traces of past channel migrations, the vertical component of which is the IHS as captured in Fig. 2b.

## 3  Modeling Approach

The workflow for modeling the point bar is as summarized in Fig. 3. This would be discussed in detail in subsequent sections.

## 4  Channel and Point Bar Facies Identification

Channel and point bar facies can be identified using well log information. Previously, Spontaneous Potential (SP) logs have been used to accomplish this task (e.g., [25, 27], where a bell shape signal has been interpreted to be a point bar while a blocky or cylindrical shape has been interpreted as a channel (Fig. 4).

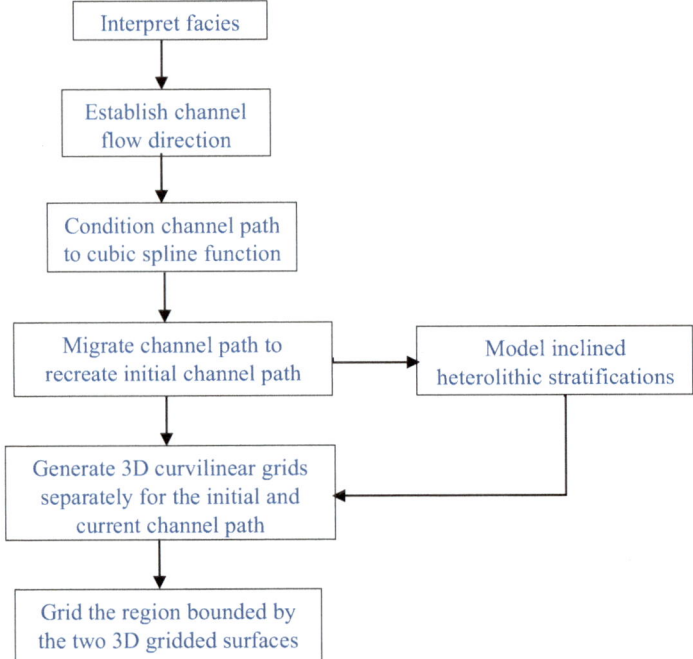

**Fig. 3** Proposed workflow for modeling the point bar reservoir

## 5   Channel Path Recreation

The workflow for channel path recreation is as summarized in Fig. 5. We begin by identifying facies and classifying them as either point bars or channels, using well log information (Fig. 5a). Since this is a synthetic workflow for demonstration purpose, we assume that all the blue points are channel well locations and the red ones are point bar locations. We then establish the direction of channel progression. This can be done either by using Gamma ray log readings, which increases in the downstream direction (because of increasing clay content), or by inferring from the variation in channel thickness, which decreases in the downstream direction [5, 28]. In this synthetic workflow, the facies have been sorted from left to right (Fig. 5b), because it has been assumed that the channel progresses easterly. The channel path recreation begins in Fig. 5c, where we honor the geological phenomenon of point bars forming at the concave side of a channel meander. This is done by conditioning the channel path to go through the channel nodes sequentially, and bend to accommodate the point bar nodes on the concave side of the bend. This process continues until the entire well data is accommodated (Fig. 5d).

The channel meander path is approximated by a parametric natural cubic spline which passes through a given sequence of channel nodes. The basic form of a cubic spline, with coefficients $a$, $b$, $c$, and $d$ is defined as:

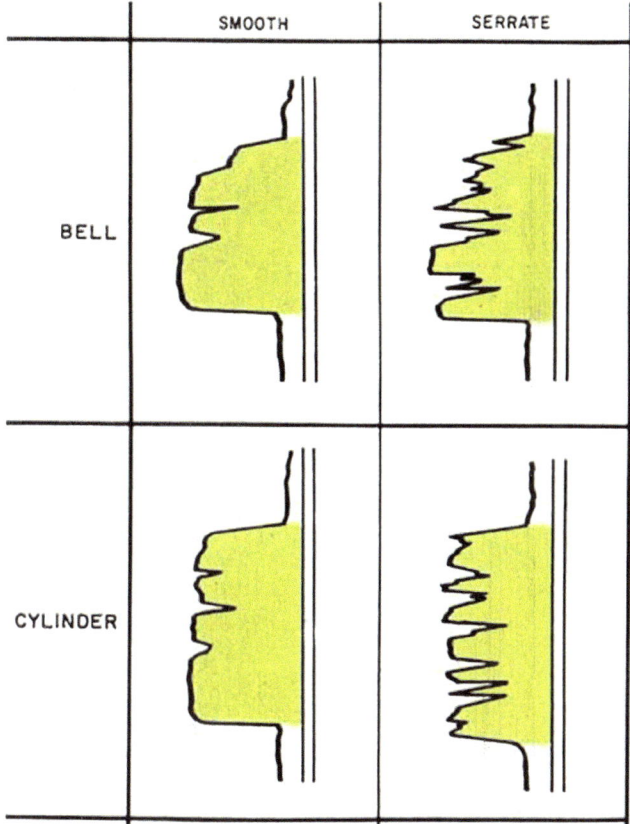

**Fig. 4** SP log profiles for point bar and channel identification. Blocky or cylindrical shape is an indication of a channel while a bell shape is an indication of a point bar. Adapted from [38]

$$P(t) = at_j^3 + bt_j^2 + ct_j + d \qquad (1)$$

The parameter value $t$ for the $jth$ channel node, denoted $t_j$ is the cumulative sum of the square root of chord length defined according to the centripetal scheme by [18], and it is expressed as:

$$t_j = \sum_{i<j} \sqrt{\|channel\ nodes_{i+1} - channel\ nodes_i\|_2} \qquad (2)$$

The coefficients $(a, b, c, d)$, which are weights for interpolating the channel nodes, could be determined from [22].

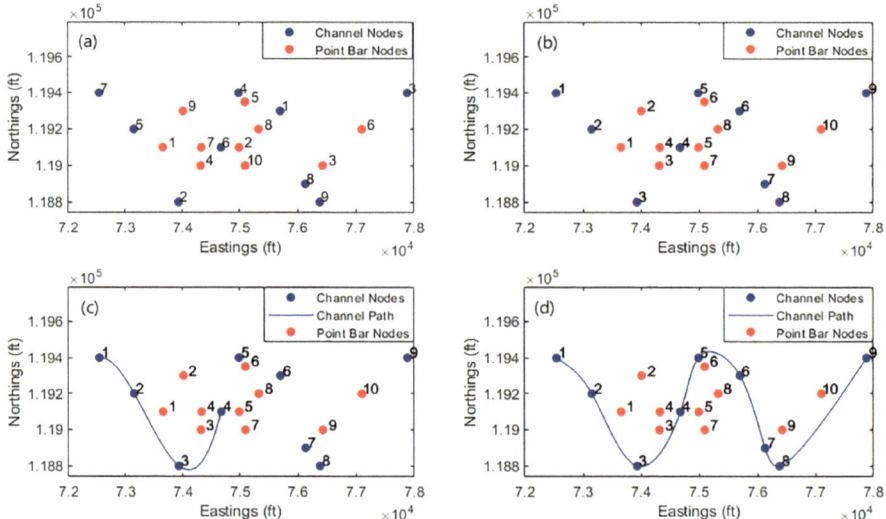

**Fig. 5** Synthetic procedure for channel path recreation. **a** Identifying facies and classifying them into channels (blue) and point bars (red), **b** Sorting facies in the direction of channel flow (East–West direction), **c** Channel path recreation begins. Channel path goes through channel nodes sequentially, and bend to accommodate the point bars on the concave side, and **d** Full channel path accommodating all the well data

## 6  Channel Path Migration

The migration process basically involves using today's channel meander path (i.e., the current channel meander path) to recreate the ancient channel path (i.e., initial channel meander path). This process allows us to capture the entire aerial extent of point bar and its lateral accretions. We can conceptually explain this process by using any of the concave regions of the channel meander path in Fig. 5d. Assuming the point bar associated with the first concave region in Fig. 5d is of interest, we isolate the channel path in that portion. This extracted channel meander path becomes the current channel path (Fig. 6a). Perturbing the spline coefficients to recreate this initial channel meander path is problematic, because ensuring that the spline coefficients exhibit consistency among themselves is extremely difficult. Instead, we accomplish the migration task by defining a focal point that controls $n$ possible channel meander paths as shown in Fig. 6b. That is, if there are $p$ points along the current channel path (with coordinates $x_i$, $y_i$, $i = 1, 2, \ldots p$, then $n$ possible points $\left(x_{ij}, y_{ij}, j = 1, 2.., n\right)$ can be generated corresponding to each of these $p$ points and the focal point $\left(x_f, y_f\right)$, using Eq. 3:

$$\left(x_{i_j}, y_{i_j}\right) = \left(x_i + (j - 1) \cdot \frac{\left(x_i - x_f\right)}{n - 1}, \; y_i + (j - 1) \cdot \frac{\left(y_i - y_f\right)}{(n - 1)}\right) \quad (3)$$

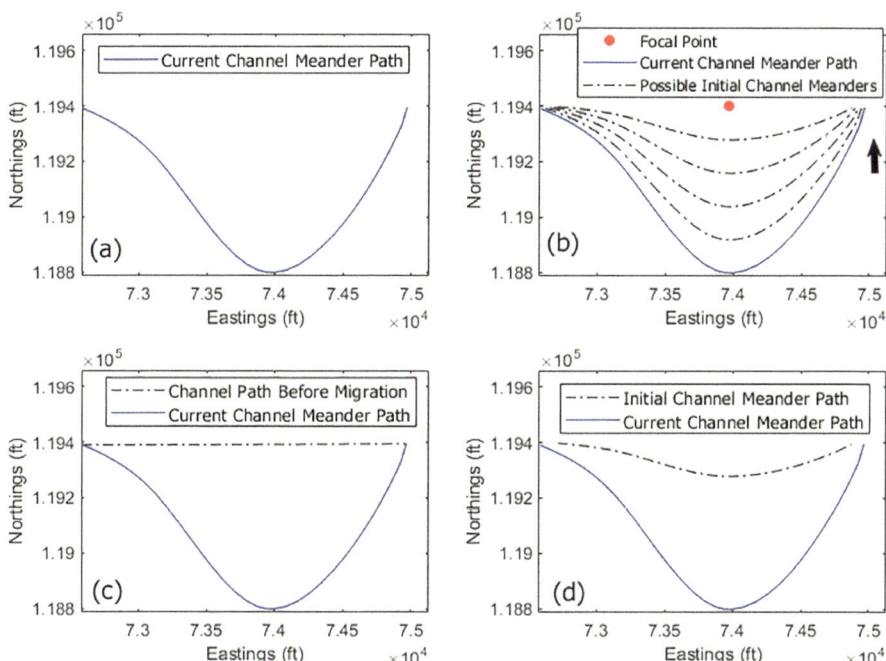

**Fig. 6** Migration of current channel path to recreate initial channel path. **a** Current channel path, **b** Current channel path migrated backwards to recreate possible initial channel meander paths, (arrow indicates the direction of backward migration, i.e., migration starting from today's channel path to the ancient path), **c** Area covered by the Current channel path and the pre-migration path and **d** Current channel path and the most probable initial channel path

where $i = 1, 2, 3 \ldots p$; $j = 1, 2, 3 \ldots n$; $x_i$ and $y_i$ are respectively, the $x$ and $y$ coordinates of each point at node $i$ along the current channel path, and $(x_f, y_f)$ are defined as:

$$
(x_f, y_f) = \begin{cases} \left( x|_{\frac{dy}{dx}=0}, \max(y_i, i = 1, \ldots, p) \right), if\ channel\ concaves\ up \\[2ex] \left( x|_{\frac{dy}{dx}=0}, \min(y_i, i = 1, \ldots, p) \right), if\ channel\ concaves\ down \end{cases} \tag{4}
$$

$x|_{\frac{dy}{dx}=0}$ is the x-coordinate at the bend where the slope of the channel path is zero. Please note that in all our discussions, it is assumed that the channel progresses in the E-W direction. In a case where the channel path is oblique, a coordinate rotation is necessary for the formulations discussed herein to work.

Applying Eq. 3 and 4 yields the focal point (red point) and the possible initial channel meander paths in Fig. 6b. To select the most probable initial meander path, we make use of the concept of erosion coefficients for each of the possible initial channel meander path. The use of erosion coefficients is guided by the observation that before lateral migration of the channel, the channel path is linear. The channel begins to bend when erosion begins to occur. Therefore, the extent of channel curvature is an indication of the degree of erosion. Thus, we can capture the extent of curvature or erosion coefficient ($\alpha$), by using the area bounded by the curves (i.e., the channel paths), as shown in Eq. 5. If knowledge or field data about the erosion co-efficient is available, we can select the initial meander path as the one that yields the closest match to the erosion coefficient from field data, after applying Eq. 5. Otherwise, one can assume equal likelihood of occurrence for each of the paths and randomly select one of the generated initial channel meander paths. In this demonstration, we used the latter to obtain the initial meander path (see Fig. 6d).

$$\alpha = 1 - \frac{A'}{A} \tag{5}$$

where $A$ is the area bounded by the pre-migration channel path and the current channel path (Fig. 6c), and $A'$ is area bounded by the initial meander path and the current channel path (Fig. 6d).

As the channel migrates, it leaves behind lateral accretion surfaces that extend from the initial channel position to the current location of the channel. The encircled portion in Fig. 7 illustrates a point bar deposit that extends up to the banks of the current channel. To capture the full extent of the point bar, we need to account for this lateral extension. If the channel has a width $W$, then the distance $d$ to which the point bar laterally extends can be computed as [2]:

$$d = \frac{W}{1.5} \tag{6}$$

As an approximation, [19] demonstrated that $W$ can be computed as:

$$W = \sqrt[1.01]{\frac{\lambda_m}{10.9}} \tag{7}$$

The parameter $\lambda_m$ is the wavelength of the channel (units in m), which is the distance between point $T$ and $K$ as illustrated in Fig. 7. In our case, it is equivalent to the length of the pre-migration path in Fig. 6c.

Using these pieces of information, we can delineate the full lateral extent of the initial and current channel meander paths, by computing the coordinates at the points of extension for the initial and current channel paths. If there are $p$ points on each channel path, then for each point $i$ on the initial or current channel path, we can compute these coordinates $\left(x_{pb,i}, y_{pb,i}\right)$, using Eq. 8.

$$\left(x_{pb,i}, y_{pb,i}\right) = \left(\left(x_{ch,i} \pm d \cdot \cos \beta_i\right), \left(y_{ch,i} \pm d \cdot \sin \beta_i\right)\right) \tag{8}$$

where $i = 1, 2, 3 \ldots p$, $\beta_i$ is the angle between a point at node $i$ on the channel path, and the focal point discussed earlier. The focal point is illustrated as $F$ in Fig. 7. The

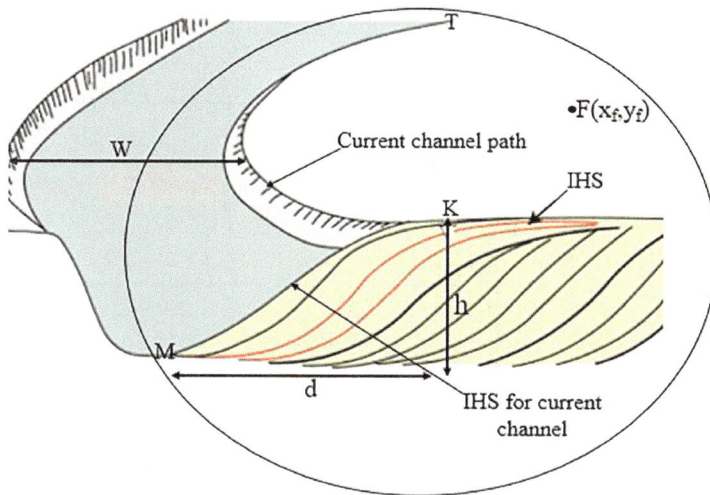

**Fig. 7** Illustration of the of point bar deposit extending into the channel, adapted from [23]

signs in Eq. 8 depend on whether $\beta_i$ is positive or negative. Applying Eq. 8, yields the path extensions for the current and initial channel meanders paths in Fig. 8.

# 7 Modeling the IHS Geometry

As described by [34], the inclined heterolithic surfaces (IHS) are approximately sigmoidal. These surfaces represent the vertical sequence of sediments that are deposited as the channel migrates. We modeled the geometry of the IHS by solving the sigmoidal equation along the IHS surfaces, using Eq. 9.

$$z° = o\frac{h}{1 + e^{-as}}$$ (9)

where $h$ is the vertical thickness of the point bar (see Fig. 7) and $a$ is the slope of an IHS over a horizontal distance $s$. As illustrated in this Fig. 7, $s = d$ for the current channel path. Applying Eq. 9 for the current channel path yields Fig. 9.

# 8 Grid Generation

The task of gridding the complex 3D geometry of a point bar was simplified somewhat by first discretizing the initial and current channel paths, and later combining them with their corresponding IHS grids, to generate 3D gridded surfaces. Finally, the

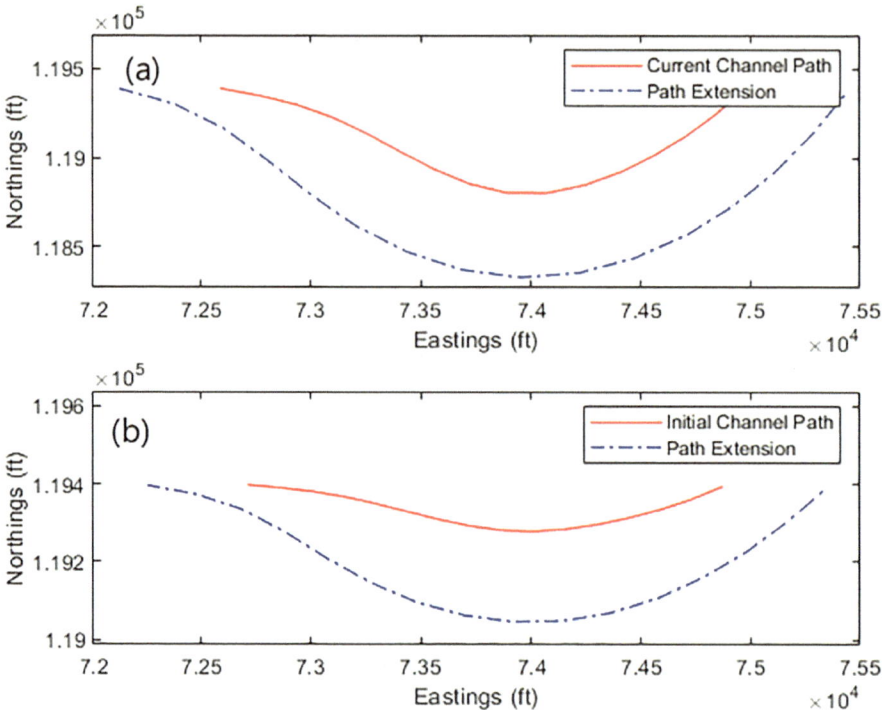

**Fig. 8** Demarcation of the lateral extents of **a** the current channel path, and **b** the initial channel path

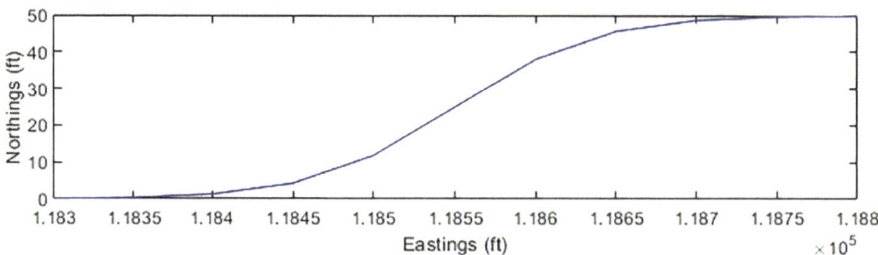

**Fig. 9** Sigmoidal representation of the IHS for the current channel path

region between the 3D gridded surfaces for the current and initial meander paths are infilled to complete the gridding of the entire point bar.

The procedure is such that, a domain of interest is initially defined, as depicted in Fig. 10a as the region between the current channel path and the migrated path. Assuming the number of grid nodes along each channel path is $nx$, and $L$ is the length along the channel, we can compute the cumulative distance $l$ at every node $i$ along the channel path using Eq. 10, and use it determine the coordinates of the grid nodes at every division along the channel path (see Fig. 10b). To generate the coordinates of the grid nodes across the channel, Eq. 11 is used by specifying the number of grid nodes across the channel $(ny)$, to produce Fig. 10c. The equivalent curvilinear grid for the current channel is displayed in Fig. 10d.

$$l_i^\circ = \circ \frac{L}{nx} \circ \cdot (i - 1) \tag{10}$$

$$\left(u_{i,j}, v_{i,j}\right) = \left(x_{i,j} + (j - 1) \cdot \frac{x_{i,j} - x'_{i,j}}{ny - 1}, y_{i,j} + (j - 1) \cdot \frac{y_{i,j} - y'_{i,j}}{ny - 1}\right) \tag{11}$$

where $i = 1, 2, 3 \ldots nx$; $j = 1, 2, 3 \ldots ny$; $\left(u_{i,j}, v_{i,j}\right)$ is the coordinate of the grid at node $i$, $j$ across the channel. $(x_{i,j}, y_{i,j})$ and $(x'_{i,j}, y'_{i,j})$ are the respective coordinates of the grid nodes generated along the channel paths (Fig. 10b).

For a point bar of constant vertical thickness, $h$, as illustrated previously, if we know the z-coordinates of the grid nodes for a section across the channel $\left(Z_{i=constant, j, k}\right)$, the remaining z-coordinates of the grid nodes at the other locations $\left(z_{i,j,k}\right)$ can be easily replicated. Thus, by specifying $nz$ (i.e., the number of grid nodes along the z-axis), we can repeat the procedure used to generate grid nodes along the channel to generate grid nodes along the IHS at a particular section (Fig. 11).

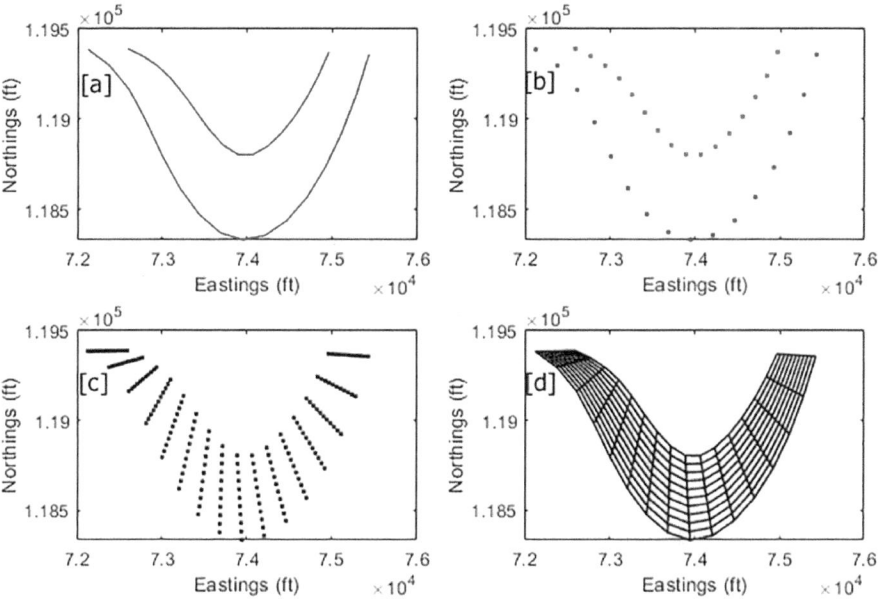

**Fig. 10.** 2D Grid generation process for the current channel path. **a** Domain to be gridded, **b** grid nodes generated along the channel, **c** grid nodes generated across the channel and **d** equivalent curvilinear grid for the current channel path

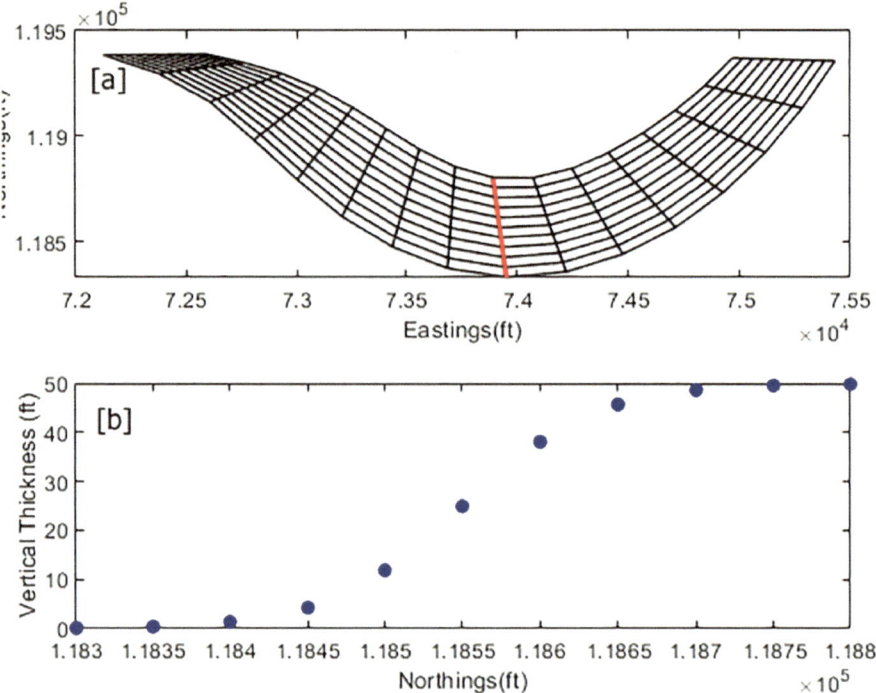

**Fig. 11  a** Current channel path, and section taken across it (red line), and **b** Grid nodes generated for the IHS across this section for the current channel path

We are now ready to project the gridded channel paths into 3D gridded surfaces. Figure 12a shows the 3D gridded surface for the current channel path. Repeating the above procedure for the initial channel path yields Fig. 12b. To generate a grid for the entire point bar, the overlap region between the two 3D gridded surfaces, as illustrated in Fig. 12c is gridded, using Eq. 12. Figure 12d represents the 3D grid for the entire point bar.

$$
\begin{aligned}
\left(X_{ijk}, X_{ijk}, Z_{ijk}\right) = \Bigg( & u_{i,j,k} + (j-1) \cdot \frac{u_{i,j,k} - u'_{i,j,k}}{ny-1}, v_{i,j,k} \\
& +(j-1) \cdot \frac{v_{i,j,k} - v'_{i,j,k}}{ny-1}, z_{i,j,k} + (j-1) \cdot \frac{z_{i,j,k} - z'_{i,j,k}}{ny-1} \Bigg)
\end{aligned}
\tag{12}
$$

where $\left(u_{i,j,k}, v_{i,j,k}, z_{i,j,k}\right)$ and $(u'_{i,j,k}, v'_{i,j,k}, z'_{i,j,k})$ are the coordinates of the grid nodes along the 3D gridded channels in Fig. 12a and b respectively.

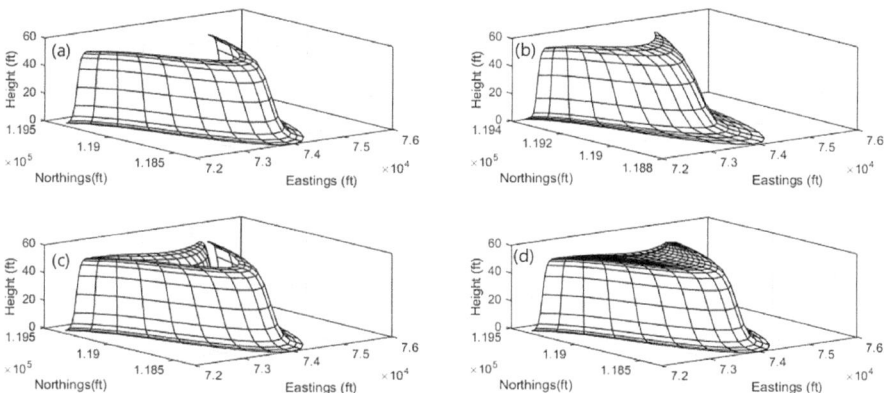

**Fig. 12.** 3D Grid generation process for the entire point bar. **a** 3D grid for the current channel, **b** 3D grid for the initial channel, **c** overlap of the 3D grids for the current and initial channel paths, showing the overlap region to be gridded **d** equivalent 3D curvilinear grid for the entire point bar

## 9 Preservation of Point Bar Architecture and Its Internal Heterogeneity

Generating a grid that preserves the point bar reservoir architecture is important, as it is critical to the preservation of the internal heterogeneities in geostatistical simulation. As can be seen in Fig. 13, horizontal (Fig. 13a) and vertical sections (Fig. 13b) taken across the point bar show that the curvilinear architecture of the reservoir is preserved by the gridding scheme implemented.

While geostatistical simulation methods like multiple point statistics (MPS) [13, 21] can offer excellent approximation of reservoir architecture and its internal properties, the grid resolution required to capture some of the fine scale variations in a point may render the MPS approach computationally burdensome. As can be seen in Fig. 13, the gridding scheme incorporated in the workflow does not necessarily require many grid cells to sufficiently approximate the point bar curvilinear geometry. The proposed approach is therefore a less computationally expensive method for modeling point bar reservoirs.

To model the point bar properties, the direct use of the conventional geostatistical simulation methods like the Sequential Gaussian Simulation [10] may yield suboptimal results. This is because these methods are implemented within a rectilinear grid system and cannot capture the curvilinear continuity of the point bar properties. Therefore, implementing a grid transformation scheme is necessary. In the grid transformation, the curvilinear grid can be transformed into an equivalent rectilinear grid within which the properties can be modeled, after which the properties can be mapped back into the original curvilinear grid. The idea is to ensure that estimates of the properties proceed in a manner that preserves the point bar reservoir heterogeneity.

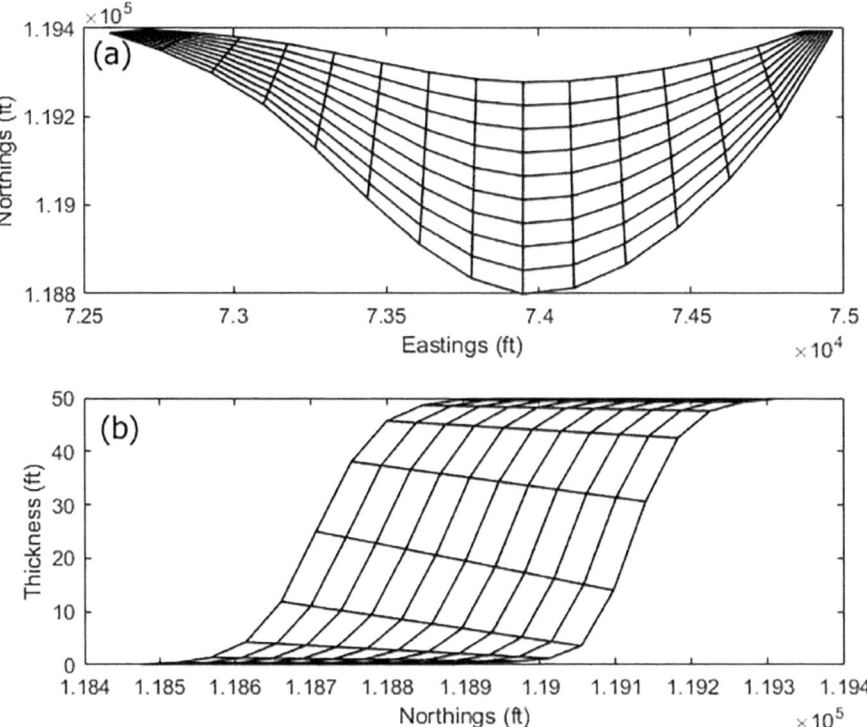

**Fig. 13** **a** horizontal slice, and **b** vertical slice taken arbitrarily across the point bar to demonstrate the preservation of the curvilinear architecture of the point bar

## 10 Concluding Remarks

A systematic method for modeling asymmetric point bar geometries has been proposed. The method incorporates a computationally inexpensive gridding scheme that accounts for the point bar curvilinear architecture. The inexpensive gridding scheme incorporated in the workflow makes the proposed method a promising technique for modeling point bars, especially when a flow simulation study is to be conducted on a large ensemble of point bar models.

## References

1. Allen, J.R.: A review of the origin and characteristics of recent alluvial sediments. Sedimentology **5**, 89–191 (1965)
2. Allen, J.R.L.: The sedimentation and paleogeography of the Old Red Sandstone of Anglesey, North Wales. Proc. Yorks. Geol. Soc. **35**, 139–185 (1965)
3. Beniot, I., Fillacier, S., Le Gallo, Y., Audigane, P., Chiaberge, C., Viseur, S.: Modelling of CO2

injection in fluvial sedimentary heterogeneous reservoirs to assess the impact of geological heterogeneities on CO2 storage capacity and performance. Energy Procedia **37**, 5181–5190 (2013)

4. Boisvert, J.B.: Conditioning object based models with gradient based optimization. **2011**(1) (2011). http://www.ccgalberta.com
5. Brierley, G.J., Hickin, E.J.: The downstream gradation of particle sizes in the Squamish River, British Columbia. Earth Surf. Proc. Land. **10**(6), 597–606 (1985)
6. Carter, D.C.: 3-D seismic geomorphology: insights into fluvial reservoir deposition and performance, Widuri field Java Sea. AAPG Bull. **87**, 909–934 (2003)
7. Clift, P.D., Olson, E.D., Lechnowskyj, A., Moran, M.G., Barbato, A., Lorenzo, J.M.: Grain-size variability within a mega-scale point-bar system, False River Louisiana. Sedimentology **66**(2), 408–434 (2019). https://doi.org/10.1111/sed.12528
8. Dawuda, I., Srinivasan, S.: A hierarchical stochastic modeling approach for representing point bar geometries and petrophysical property variations. Comput. Geosci. **164**, 105127 (2022). https://doi.org/10.1016/j.cageo.2022.105127
9. Deutsch, C.V.: A sequential indicator simulation program for categorical variables with point and block data: BlockSIS. Comput. Geosci. **32**(10), 1669–1681 (2006). https://doi.org/10.1016/j.cageo.2006.03.005
10. Deutsch, C.V., Journel, A.G.: GSLIB: Geostatistical Software Library and User's Guide (Second). Oxford University Press, New York (1997)
11. Deutsch, C.V., Tran, T.T.: FLUVSIM : a program for object-based stochastic modeling of fluvial depositional systems $. **28**, 525–535 (2002)
12. Durkin, P., Hubbard, S.M., Boyd, R.L., Leckie, D.A.: Stratigraphic expression of intra-point-bar erosion and rotation. J. Sediment. Res. **85**, 1238–1257 (2015)
13. Eskandari, K., Srinivasan, S.: Reservoir modelling of complex geological systems—a multiple point perspective. Can. Int. Pet. Conf. **2008**, 59–68 (2010). https://doi.org/10.2118/2008-176
14. Fielding, C.R., Crane, R.C.: An application of statistical modelling to the predicton of hydrocarbon recovery factors in fluvial reservoir sequences. In: Ethridge, F.G., Flores, R.M., Harvey, M.D. (eds.) Recent Developments in Fluvial Sedimentology. Special Publications, vol. 39 (1987)
15. Güneralp, İ, Marston, R.A.: Process–form linkages in meander morphodynamics: bridging theoretical modeling and real world complexity. Prog. Phys. Geogr. **36**(6), 718–746 (2012)
16. Issautier, B., Fillacier, S., Le, Y., Audigane, P., Chiaberge, C., Viseur, S.: Modelling of CO2 injection in fluvial sedimentary heterogeneous reservoirs to assess the impact of geological heterogeneities on CO2 storage capacity and performance. Energy Procedia **37**, 5181–5190 (2013). https://doi.org/10.1016/j.egypro.2013.06.434
17. Issautier, B., Viseur, S., Audigane, P., le Nindre, Y.M.: Impacts of fluvial reservoir heterogeneity on connectivity: implications in estimating geological storage capacity for CO2. Int. J. Greenhouse Gas Control **20**, 333–349 (2014). https://doi.org/10.1016/j.ijggc.2013.11.009
18. Lee, E.T.Y.: Choosing nodes in parametric curve interpolation. Comput. Aided Des. **21**(6), 363–370 (1989). https://doi.org/10.1016/0010-4485(89)90003-1
19. Leopold, L.B., Wolman, M.G.: River meanders. Geol. Soc. Am. Bull. **71**, 789–794 (1960)
20. Mackey, S.D., Bridge, J.S.: Three-dimensional model of alluvial stratigraphy; theory and applications. J. Sediment. Res. **65**(1b), 7–31 (1995)
21. Mariethoz, G., Caers, J.: Multiple-point geostatistics: stochastic modeling with training images (2014)
22. McKinley, S., Levine, M.: Cubic spline interpolation. **45**(1), 1049–1060 (1998)
23. McMahon, W.J., Davies, N.S.: The shortage of geological evidence for pre-vegetation meandering rivers. In Fluvial Meanders and Their Sedimentary Products in the Rock Record (2018). https://doi.org/10.1002/9781119424437.ch5
24. Miall, A.D.: Reconstructing the architecture and sequence stratigraphy of the preserved fluvial record as a tool for reservoir development: a reality check. AAPG Bull. **90**, 989–1002 (2006)
25. Nazeer, A., Abbasi, S.A., Solangi, S.H.: Sedimentary facies interpretation of Gamma Ray (GR) log as basic well logs in Central and Lower Indus Basin of Pakistan. Geod. Geodyn. **7**(6), 432–443 (2016). https://doi.org/10.1016/j.geog.2016.06.006

26. Niu, B., Bao, Z., Yu, D., Zhang, C., Long, M., Su, J., Gao, X., Zhang, L., Zang, D., Li, M., Li, Y.: Hierarchical modeling method based on multilevel architecture surface restriction and its application in point-bar internal architecture of a complex meandering river. J. Petrol. Sci. Eng. **205**(April), 108808 (2021). https://doi.org/10.1016/j.petrol.2021.108808

27. Odundun, O., Nton, M.: Facies interpretation from well logs: applied to SMEKS field, offshore Western Niger Delta. Am. Assoc. Pet. Geol. **25** (2011)

28. Pitlick, J., Cress, R.: Downstream changes in the channel geometry of a large gravel bed river. Water Resour. Res. **38**(10), 34–41 (2002)

29. Pyrcz, M.J., Boisvert, J., Deutsch, C.V.: ALLUVSIM : a program for event-based stochastic modeling of fluvial depositional systems $. Comput. Geosci. **35**, 1671–1685 (2009). https://doi.org/10.1016/j.cageo.2008.09.012

30. Pyrcz, M.J., Catuneanu, O., Deutsch, C.V.: Stochastic surface-based modeling of turbidite lobes. Am. Asso. Petrol. Geol. Bull. **89**(2), 177–191 (2005). https://doi.org/10.1306/092204 03112

31. Pyrcz, M.J., Deutsch, C.V.: Stochastic simulation of inclined heterolithic stratification with streamline-based stochastic models. In: Center for Computational Geostatistics Annual Report Papers, 1–14. papers2://publication/uuid/4BADF3C6-381A-499D-A4C5-DD63447E3CAE (2004)

32. Shu, X., Hu, Y., Jin, B., Dong, R., Zhou, H., Wang, J.: Modeling method of point bar internal architecture of meandering river reservoir based on meander migration process inversion algorithm and virtual geo-surfaces automatic fitting technology. SPE Annu. Tech. Conf. Exhib. (2015). https://doi.org/10.2118/175013-MS

33. Su, Y., Wang, J.Y., Gates, I.D.: SAGD well orientation in point bar oil sand deposit affects performance. Eng. Geol. **157**, 79–92 (2013). https://doi.org/10.1016/j.enggeo.2013.01.019

34. Thomas, R.G., Smith, D.G., Wood, J.M., Visser, J., Calverley-Range, E.A., Koster, E.H.: Inclined heterolithic stratification-terminology, description, interpretation and significance. Sed. Geol. **53**, 123–179 (1987)

35. Werren, E.G., Shew, R.D., Adams, E.R., Stancliffe, R.J.: Meander-belt reservoir geology, middip Tuscaloosa, Little Creek field, Mississippi. In: Sandstone Petroleum Reservoirs. Springer, New York, NY (1990)

36. Wightman, D.M., Pemberton, S.G.: The lower cretaceous (Aptian) McMurray formation: an overview of the McMurray area, northeastern Albert. In: Pemberton, G.S., James, D.P. (eds.) Petroleum Geology of the Cretaceous Lower Manville Group: Western Canada. Can. Soc. Pet. Geol. **18**, 312–344 (1997)

37. Willis, B.J., Tang, H.: Three-dimensional connectivity of point-bar deposits. J. Sediment. Res. **80**(5–6), 440–454 (2010). https://doi.org/10.2110/jsr.2010.046

38. Wilson, B.W., Nanz, R.H.: Sand conditions as indicated by the self-potential Log. In: *EPRM Memorandum Report* (1959)

39. Yin, Y.: A new stochastic modeling of 3-D mud drapes inside point bar sands in meandering river deposits. Nat. Resour. Res. **22**(4), 311–320 (2013). https://doi.org/10.1007/s11053-013-9219-3

# Application of Reinforcement Learning for Well Location Optimization

**Kshitij Dawar⊙, Sanjay Srinivasan⊙, and Mort D. Webster⊙**

**Abstract** The extensive deployment of sensors in oilfield operation and management has led to the collection of vast amounts of data, which in turn has enabled the use of machine learning models to improve decision-making. One of the prime applications of data-based decision-making is the identification of optimum well locations for hydrocarbon recovery. This task is made difficult by the relative lack of high-fidelity data regarding the subsurface to develop precise models in support of decision-making. Each well placement decision not only affects eventual recovery but also the decisions affecting future wells. Hence, there exists a tradeoff between recovery maximization and information gain. Existing methodologies for placement of wells during the early phases of reservoir development fail to take an abiding view of maximizing reservoir profitability, instead focusing on short-term gains. While improvements in drilling technologies have dramatically lowered the costs of producing hydrocarbon from prospects and resulted in very efficient drilling operations, these advancements have led to sub-optimal and haphazard placement of wells. This can lead to considerable number of unprofitable wells being drilled which, during periods of low oil and gas prices, can be detrimental for a company's solvency. The goal of the research is to present a methodology that builds machine learning models, integrating geostatistics and reservoir flow dynamics, to determine optimum future well locations for maximizing reservoir recovery. A deep reinforcement learning (DRL) framework has been proposed to address the issue of long-horizon decision-making. The DRL reservoir agent employs intelligent sampling and utilizes a reward framework that is based on geostatistical and flow simulations. The implemented approach provides opportunities to insert expert information while basing well placement decisions on data collected from seismic data and prior well tests. Effects of prior information on the well placement decisions are explored and the developed DRL derived policies are compared to single-stage optimization methods for reservoir development. Under similar reward framework, sequential well placement strategies developed using DRL have been shown to perform better than simultaneous drilling of several wells.

K. Dawar (✉) · S. Srinivasan · M. D. Webster
The Pennsylvania State University, State College, PA 16801, USA
e-mail: sebastian.avalos@queensu.ca; kshitij.dawar@gmail.com

© The Author(s) 2023
S. A. Avalos Sotomayor et al. (eds.), *Geostatistics Toronto 2021*, Springer Proceedings in Earth and Environmental Sciences, https://doi.org/10.1007/978-3-031-19845-8_7

**Keywords** Deep reinforcement learning · Well location optimization · Markov decision process

# 1 Introduction

In recent decades, there has been a renewed interest in the decision-making process for optimal well placement. The reasons for this are twofold. First, there has been a proliferation in the amount of data collected during reservoir development that can help guide the decision-making process. Second, improvement in transistor technology has enabled the use of artificial intelligence algorithms for utilizing data collected to maximize recovery. These improvements in computer technology have led to faster and computationally cheaper flow simulations leading to the consideration of various scenarios that represent uncertainty in underlying reservoir properties. In addition, the faster computers also facilitate the training process that is intrinsic to most machine learning methods.

Due to improvements in horizontal drilling technologies, several unconventional reservoir plays have been rendered profitable. The optimal method to maximize economic return from these assets is through the strategic placement of wells in high productivity zones, minimizing the number of wells required for the efficient recovery of hydrocarbons from the reservoir. Optimal well placement also emphasizes strategic extraction of knowledge about the subsurface which in turn promotes the development of superior reservoir models, reducing the uncertainty in future well placement decisions.

Early methods for well location optimization focused on considering wide varieties of constraints for adjoint or mixed-integer optimization. These constraints would include geologic uncertainties, cost estimates, fluid properties, facilities etc. Methods such as mixed-integer programming have been used to optimize the placement of well locations [1–3] by expressing the objective function as a linear combination of the variables. These methods have the advantage of being fast and give an easily interpretable solutions to the problem. But these optimization techniques fail to characterize the highly non-linear relationships between reservoir variables, and between these variables and the dynamic response. Gradient-based optimization methods allow for the inclusion of non-linear relationships between variable and rely on the computation of the gradient of a prespecified optimization function. The adjoint-based formulations for gradient-based optimization allows for the simpler interpretation of the constraints and allows for the identification and isolation of the interesting well regions in the reservoir [4]. It has the added advantage of leading to faster convergence as compared to other gradient methods while being highly versatile and interpretable. These methods allow for the inclusion of diverse decisions variables and are usually combined with conventional flow simulation. As the well placement problem is inherently a dynamic programming problem, any increase in the range of decision-variables leads to an exponential increase in the range of choices that need to be considered for the placement problem. This also leads to an increase in the number of flow simulations that need to be considered for the optimization problem. Due to the computational costs, these methods utilize ingenious approaches to reduce the search space, but they are heavily reliant upon the definition of the adjoint equation.

One of the most popular methods for well location optimization is the use of population-based optimization algorithms, more specifically Genetic algorithms (GA). GA based optimization aims to replicate the process of natural selection in a control environment. They function by initializing a population of candidate well locations and evaluating the value of the objective function (also known as determining the fitness of the population). The 'fittest' wells are then selected for the reproduction step in which new wells will be determined by utilizing 'mutations' and 'crossovers'. Bittencourt and Horne [5] introduced GA to the oil and gas domain for the purpose of well location optimization. Their approach aimed to reduce the search space by using Tabu search [6, 7]. To include geostatistical inputs into the GA formation, several authors [8, 9] have included kriging into their framework. This helped improve the interpretability of GA results at the cost of increased computational requirement. Other authors have conducted studies into the explanatory power of GA and on the effect of hyper-parameters [10], extensions to horizontal wells [11] and gas-condensates [11, 12] etc. GA-based algorithms rely on sampling-based optimization, but the approach is plagued by issues, such as, non-optimality of solutions, vast number of hyper-parameters, high computational costs for evaluating candidate well locations.

Metrics such as productivity potential maps [13] can be extremely useful in speeding up the evaluation of the well locations. In addition, research into use of neural networks for forecasting production [14] and developing earth models [15, 16] can be included in the proxy model formulation to develop a holistic view of the uncertainty in the reservoir properties while doing away with the need for full flow simulations. These deep learning models enable the use of transfer learning [17] accelerating the training process. In addition, these transfer learning models can be built independently by private oil and gas companies in conformance to their policy of data privacy.

The field of reinforcement learning is uniquely positioned to address dynamic programming problems. With recent advancements in the use of deep learning for functional approximations of reinforcement learning solutions, the applicability of the DRL methods have skyrocketed. Some of the popular uses include playing chess [18], controlling robots [19], improving cybersecurity [20] etc. A key insight into the use of reinforcement learning for well location optimization is that these methods have been extensively utilized for addressing multi-stage decision-making under uncertainty. Recent application of reinforcement learning to the field of geosciences include slope stability analysis for landslide prevention [21] and for determination of the first arrival for seismic image processing [22]. Extensive research into improving the sampling [23], memory buffers [24], addressing biases [25, 26] etc. has led to improved applicability of DRL for problems previously perceived to be too large and too complex to be addressed by reinforcement learning.

Here, we show the application of reinforcement learning methods to the well location problem. In Sect. 2, we take an in-depth look into the theory of the reinforcement learning algorithms applied in the research. Section 3 discusses the well location problem and highlights the applicability of the reinforcement learning approach to addressing the problem. Section 4 demonstrates the application of reinforcement learning to two case studies. The first case studies address the optimization of wells in a 2-D reservoir while Case 2 addresses the optimization of a 3-D reservoir. Sections 5 and 6 focus on the discussions on the case studies and provide concluding remarks.

## 2  Theory

Reinforcement learning involves an agent (or a decision maker) interacting with its environment $\mathcal{E}$, usually formulated as a Markov decision process $(\mathcal{S}, \mathcal{A}, T, r, \gamma)$ [27] with state space $\mathcal{S}$, action space $\mathcal{A}$, reward $r$, transition function $T(s'|s, a)$ and discount factor $\gamma \in [0, 1]$. At each time step $t$, the agent takes an action $a \in \mathcal{A}$ in state $s \in \mathcal{S}$ and transitions to state $s'$ while receiving a reward $r(s, a, s') \in \mathbb{R}$. The goal of the agent is the maximize the expected cumulative discounted future reward, or return, $R_t = \sum_{t'=t}^{T} \gamma^{t'-t} r_t$, where $T$ is the time-step at termination. The maximum expected return achievable by following any strategy, after being in state $s$, and then taking some action $a$ is defined as the optimal action-value function $Q^*(s, a)$. It can be mathematically represented as,

$$Q^*(s, a) = \max_{\pi} \mathbb{E}[R_t | s_t = s, a_t = a, \pi] \tag{1}$$

where, $\pi : \mathcal{S} \to \mathcal{A}$ is a policy mapping states to actions. It defines the mechanism by which the agent selects an action at state $s$. If $Q^*(s', a')$ has been determined for all possible action $a'$, then the optimal policy would be to select the $a'$ that would maximize the expected value of the future reward, $r + \gamma Q^*(s', a')$. The optimal action-value function obeys the Bellman equation [28].

$$Q^*(s, a) = \mathbb{E}_{s'} \left[ r + \gamma \max_{a'} Q^*\left(s', a'\right) | s, a \right] \tag{2}$$

The $\gamma \max_{a'} Q^*(s', a')$ term in the equation highlights the consideration of future actions in future states on the determination of the value of the current action.

$$\max_{s', a'} Q^*\left(s', a'\right) \geq 0 \to \mathbb{E}_{s'} \left[ r + \gamma \max_{a'} Q^*\left(s', a'\right) | s, a \right] \geq \mathbb{E}[r] \tag{3}$$

This differentiates DRL from conventionally employed optimization techniques which instead focus on the maximization of immediate expected reward. We can then determine the action-value function iteratively, $Q_{i+1} =$

$\mathbb{E}\left[r + \gamma \max_{a'} Q_i(s', a') | s, a\right]$. When the action-value function converges to $Q^*$, the optimal policy can then be determined, $\pi^* = \operatorname{argmax}_{a'} Q^*(s', a')$. This convergence usually takes place when the state-action space has been thoroughly sampled. One way to determine a new policy from an action-value function is to act $\epsilon$-greedy with respect to the action-values i.e., taking an action with the highest action value with a probability of $1 - \epsilon$, and taking a random action with a probability of $\epsilon$. By introducing $\epsilon$ component, exploration is introduced i.e., a sub-optimal action-value may be selected with the objective of sampling the action-value space more broadly. The expectation is that the agent will then be able to improve upon its value function estimates. During the initial training of the DRL algorithm, a high $\epsilon$ allows for better exploration of the state-action space. As the RL algorithm learns more about the environment, a lower $\epsilon$ would allow for fine-scale refinement to the action-value function. Q-learning, a form of temporal-difference (TD) [29] learning, is frequently used to estimate the optimal action values. Usually, due to the computational requirements associated with sampling each state-action pair, functional approximators are utilized to estimate the action-value function, e.g., Deep Q-Network (DQN) algorithm [30] uses neural networks as functional approximators. The action-value function is approximated using the neural network (or Q-network) with parameters $\theta$. The Q-network is trained by minimizing the loss function $L_i(\theta_i)$ at every iteration $i$, described by the following equation:

$$L_i(\theta_i) = \mathbb{E}_{s,a}\left[\left(\mathbb{E}_{s'}\left[r + \gamma \max_{a'} Q(s', a'; \theta_{i-1} | s, a)\right] - Q(s, a; \theta_i)\right)^2\right] \quad (4)$$

To optimize the performance of the neural network, mini batch stochastic gradient descent can be conducted [31]. Under this approach, the NN parameters are updated considering a stochastic approximation (using vectorization) of the gradient of the loss function. To improve convergence with the functional approximator, an experience replay memory can be generated in which agent's experience $(s_t, a_t, r_t, s_{t+1})$ are stored at every time step. The Q-learning updates are then conducted using random mini-batch samples from the memory. The mechanism for the update of the neural network parameters $\theta$ is given as:

$$\Delta\theta = \alpha\left[r + \gamma \max_{a'} Q(s', a'; \theta')\right] \nabla_\theta Q(s, a; \theta) \quad (5)$$

The target in the loss function is dependent on the trained neural network weights from the previous iteration and can change over multiple iterations. Hence, as the neural network minimizes the loss after every training step, there is an update to the neural network parameters $\theta$ which leads to the target, $Q(s, a; \theta_i)$, changing or 'moving' as well. This problem of moving targets can be addressed by dissociating the training of the target and the estimate by using separate neural networks. In the above equation, the learning rate, $\alpha$, controls the change in the model weights to the estimated error. Hyperparameter tuning is conducted to control the learning rate, but, usually, a smaller $\alpha$ leads to convergence to optimal weights, this comes at the cost of longer training time.

Another issue with the DQN formulation is the overestimation bias in the maximization step. This can lead to poor learning. One way this issue can be addressed is by decoupling the selection of the action from its evaluation. This is known as Double DQN or DDQN [25] and the new target is,

$$Y_i = r + \gamma Q\left(s', \underset{a'}{\mathrm{argmax}} Q\left(s', a'; \theta_i\right); \theta_i'\right) \tag{6}$$

The selection of the action is due to the online weights $\theta_i$ and the second set of weights $\theta_i'$ is used to evaluate the value of the policy. The weights $\theta_i'$ are updated by switching the roles of the target and the value network. The updates to the weights $\theta_i'$ are conducted after every $\tau$ steps to prevent the problem of moving targets.

## 3   Well Location Problem

During the early phases of reservoir development minimal information regarding the reservoir properties is available leading to decision-making for reservoir development under uncertainty. This problem of placing wells sequentially in an uncertain environment can be formulates as be a Markov decision process. A reservoir DRL agent can be trained to solve the problem of the optimal exploitation of the reservoir resources.

In the early phases of reservoir development, well data is not available and seismic data forms the basis for any decision-making. Existing geostatistical techniques such as stochastic simulation can be utilized to generate multiple reservoir realizations using seismic data as secondary data. These reservoir realizations reflect the uncertainty in the reservoir properties that are simulated conditioned to little to no hard data. Also, in many cases the semi-variogram inferred and modeled based on the secondary data does not accurately represent the spatial trends in the reservoir property being simulated. In these situations, inputs from geologists, petro-physicists and reservoir engineers are critical for accurately identifying trends in reservoir properties. Usually, the reservoir models guide the well placement decisions and the additional information derived, after conducting well surveys, is used to update the reservoir models. With additional data regarding the reservoir becomes available, the uncertainty in reservoir properties reduce which in turn allows for better decision-making.

This forms the framework for training the reservoir DRL agent (see algorithm in [25]). The agent starts in an unexplored state, i.e., in a state where no well data is available, with the goal of maximizing the cumulative hydrocarbon production after the placement of $N$ wells. This ensures that the problem is of the fixed-horizon type i.e., the end-state is well defined. At any given time, the reservoir DRL agent can act by placing a well at a valid location. After the action selection, the reservoir DRL agent receives information regarding the petrophysical properties at the drilled point and a reward reinforcing the goodness of the action selected.

The paper presents a simplified approach for addressing uncertainty in reservoir properties by focusing on the lithofacies as an indicator of the goodness of a selected action. The reason for this is two-fold: first, there have been several studies conducted emphasizing the relationship between lithofacies and other petrophysical properties such as porosity and permeability [32–38]; second, categorical variables like lithofacies can dramatically reduce the state-space in the DRL formulation. The reservoir DRL agent attempts to place wells in high productivity regions and balances the additional information gained by placement of a well against the economic profitability due to the hydrocarbon resources extracted from the selected regions.

Several case studies have been conducted to test the efficacy of the recommended policies and the robustness of the policies to initial assumptions about the reservoir. The goal of the DRL agent is not to present the best policy for all given combinations of assumptions regarding the geo-spatial distributions of the lithofacies. Training an DRL agent requires several assumptions regarding the development of reservoir models and the role that experts in the field of geology, petroleum engineering and earth science play in the decision-making process. Errors in the decisions taken to build the ensemble of reservoir models will affect the developed policies.

# 4 Case Studies

Case 1A focuses on the application of DRL for determination of optimal policy using Double Deep Q Network (Double DQN or DDQN). Cases 1B, 1C and 1C_alternate focus on the applicability of the developed policy in cases where the initial assumptions regarding the reservoir were flawed. The neural network architecture and the hyperparameters for the simulations can be found in the appendix. The process flow is demonstrated in Fig. 1 considering various pathways for simulating the environment. The update to the prior probabilities of pay-facies is crucial to the determination of the optimal policy. There are several methodologies to conducting this update.

1. Using fast-variogram computations and generating an updated ensemble of reservoir realizations. This method is the slowest but would yield the most statistically accurate updates.
2. Using initial ensemble to derive correlation between locations and using data assimilation tools, such as Ensemble-based Kalman filtering (EnKF) to update reservoir realizations. This method requires some preprocessing to derive necessary statistics and relies on a multi-Gaussian assumption.
3. Using individual realizations for DRL training in each episode. This is the most computationally efficient method of training the DRL agent. This presumes that the initial ensemble has within it a set of models that are consistent with the extracted new information. Thus, updates of the initial ensemble of reservoir realizations would be not required. This is the most computationally efficient

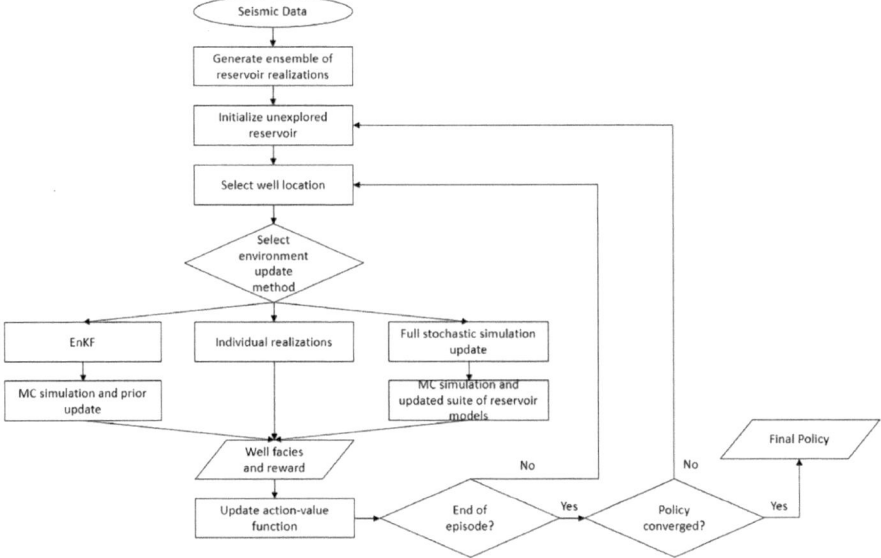

**Fig. 1** Workflow for determination of optimal policy for exploiting the reservoir using reinforcement learning

method for deriving the policy though it relies on the exhaustive characteristics of the initial set of reservoir realizations. This approach also allows for the parallelization of the learning process (Table 1).

For the case studies, the dynamics of the environment is simulated by considering a randomly selected reservoir realization for every episode (one pass through the training process, from the initial unexplored state to the terminal state) from the

**Table 1** Single stage optimization algorithm

Place first well in location with highest ensemble pixel-wise average (maximum probability) of pay-facies;
**for** *wells = 2 to M* **do**
  $A$ = locations where, pixel-wise average of pay facies at location =
    Max ((pixel-wise average of pay-facies from remaining possible well locations)-
    (well constraint penalties));
  **if** card(A)=1, **then**
    Place well in location $A$;
  **else**
    Place well in location $A(i)$ where sum of pixel-wise average of pay-facies of
    grids surrounding $A(i)$ is maximum card(A)=1 $i$ card(A)=1 $A$;
  **end**
**end**
Conduct gradient descent by perturbing well locations until converges

initial ensemble generated using seismic data. Using a lookup table (with the facies values stored for the selected realization) allows for fast computation of the next state and reward that is delivered to the reservoir DRL agent. The encoding of the state and the determination of the reward function are vital to the implementation of the deep reinforcement learning algorithm. The state encodes new information gained by placing a well in a reservoir. This information can be well log data, production data, core analysis information etc. Information regarding the lithofacies at the drilled location has been considered for the case studies presented here. This information can be encoded as a vector in two ways.

1. As a vector consisting of facies information at all grid points including place-holders for grid points where the information is not available. This formulation is efficient in cases where the reservoir is discretized into few grid points.

$$s_t = (f_1, f_2, \ldots, f_n) \tag{7}$$

where $f_i$ represents the facies at the $i$th grid point.

2. As a vector consisting of well location and determined facies at drilled well locations. This formulation works better for large reservoirs discretized into several grid blocks.

$$s_t = (x_1, x_2, \ldots, x_N, y_1, y_2, \ldots, y_N, f_1, f_2, \ldots, f_N) \tag{8}$$

where $x_i$, $y_i$ and $f_i$ represent the x-coordinate, y-coordinate, and facies at the $i$th grid point.

The latter formulation has been considered for the case studies. Recent research into deconvolutional neural networks [39] have shown their efficacy for the extension of point information in multi-dimensions to regions where the information is unavailable and/or uncertain [40, 41]. Though this has not been explored in the current research, the authors aim to explore this in future work.

The agent receives a reward determining the desirability of placing the well at the grid location and as described earlier the reservoir DRL agent's sole goal is to maximize the cumulative expected reward over $N$ well placement decisions. The definition of the reward function dictates the key criterion that the reservoir DRL agent will focus upon to optimize the well location. Though running full physics flow simulations to compute the reward is most accurate with perfect information regarding the subsurface, it is computationally infeasible after the placement of every single new well in an episode. Hence, a proxy linear regression model has been developed to return a reward to the reinforcement learning agent.

Several factors affect the desirability of well location, such as the facies at the well location, facies of surrounding grid blocks, distance to other wells, distance to the reservoir boundary etc. The proxy model was developed by conducting flow simulations for several reservoir models keeping the fluid properties constant and varying the prespecified parameters. The correlation between the variables and the target variable, the total hydrocarbon output, has been shown in Fig. 2.

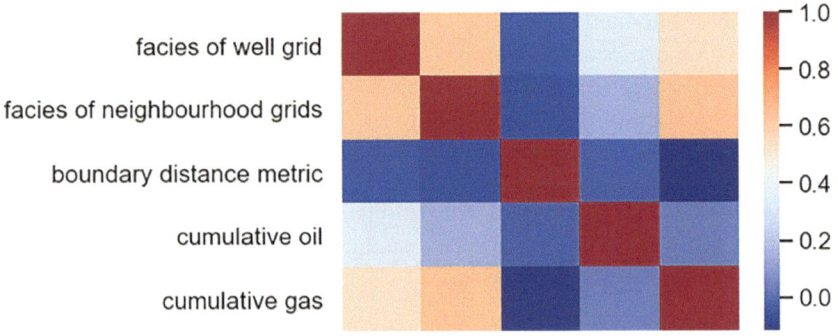

**Fig. 2** Correlation heat map between well location and facies connectivity parameters and well production

The following regression equation is used as a proxy for the flow.

$$C_0 + C_{w_{fac}} X_{w_{fac}} + C_{nh_{fac}} X_{nh_{fac}} + C_b X_b + C_{w_{PPM}} X_{w_{PPM}} = reward \qquad (9)$$

$w_{fac}$ = facies of grid at well location,
$nh_{fac}$ = facies of grids in the neighborhood grids,
$b$ = distance to the reservoir boundary,
$w_{PPM}$ = productivity potential metric to account for well spacing.

It is to be noted that the developed proxy model assumes a linear relationship between the variables. This allows for the faster computation of the reward function. Non-linear reward functions accounting for the non-linear relationships between reservoir properties and fluid production will provide more refined reward results and such relationships can be accounted for with the use of deep learning models, though this has been left for future research. Also, the reward function lends to the addition of expert information regarding the reservoir into the reinforcement learning framework. Expert information can be incorporated into the reward function by modifying the relationships between the input variables (facies in the case studies considered in this paper) and the reward derived, for e.g., well pattern constraints can be enforced by penalizing wells that are not in desired pattern.

The effect of overlapping influence of neighborhood wells on the production of a placed well is included in the proxy model in the form of Productivity Potential Maps (PPM). The regression equation for reward is developed by conducting flow simulations considering various well locations across various reservoir realizations conditioned to the initial seismic data. Reinforcement learning based methods allow for infinite customization of the optimization of well location through the modification of the state definition, action space and most importantly the reward function. In this research, we have implemented the $\epsilon$-greedy policy which involves the selection of the action associated with the highest action-value function with a probability of $1 - \epsilon$ and a random action with a probability of $\epsilon$. Other choices for the exploration (or behavioral) policy include Boltzmann policy, Boltzmann Gumbel policy [42], SoftMax policy [43] etc. Our selection of the $\epsilon$-greedy policy is based on its ease of application and low memory requirements as compared to other choices of behavioral policy.

The reward function optimization of a 2-D reservoir needs to account for information from surrounding grid points, but the formulation of the state vector includes the facies and coordinates of the well location only. For the optimization of a 3-D reservoir, the state formulation needs to account for the facies of individual grid points across the vertical section of the reservoir as a simple aggregation of the facies along the well path will lead to loss of vital information regarding the distribution of facies across the horizontal layers and the correlation of facies between reservoir layers. Another point to consider is the greater variety of potential well paths that can be drilled in a 3-D reservoir (inclined and lateral wells), but the inclusion of non-vertical well paths leads to a combinatorial explosion and has been discussed in [44] but is not included here for the sake of brevity.

The choice of number of decision-stages can also have an adverse effect on the computational time. In case of single well placement decision-stages, the RL problem will have to account for the value function of $nPk$ grid points where $k$ is the total number of wells and $n$ is the total number of possible initial well locations. For the case studies considered, the number of wells selected has been set to 5 (placed in sequential decision-making stages) which allows for an interesting comparison of DRL with single-stage optimization techniques while being computationally efficient.

Case 1A considers a 2D reservoir ($25 \times 25$ grid points) shown in Fig. 3 with initial seismic impedance map shown in Fig. 4. The regions in red represent pay facies regions (encoded as 1 in the DRL state function) and the ones in purple represent the non-pay facies (encoded as 0). As evident from the figures, the base reservoir realization has 3 distinct pay facies regions (represented by the three red bands). The upper channel is not well represented in the seismic impedance map while the lower 2 channels are slightly displaced in the seismic map. The issue of lack of precision of seismic data is frequently encountered in reservoir characterization and reservoir modeling methods need to be able to identify such translations and diffusions of features.

The goal of the reservoir DRL agent is to maximize the cumulative rewards for the placement of 5 wells sequentially. An initial ensemble of 1000 reservoir realizations is generated using sequential indicator simulation using locally varying mean by utilizing the seismic data as secondary data. The size of the ensemble used to train the agent is a compromise between processing time and realized reward (as discussed in [44]). Recognizing that and adequate ensemble is necessary to reach an optimum reward, we selected an ensemble size of 1000 realizations. The suite of reservoir models generated are unconditional (no hard data is available) but, importantly, the reservoir realizations generated adhere to the distribution of pay facies probabilities developed using the seismic impedance maps. The reservoir realizations can

**Fig. 3** Original reservoir lithofacies for Case 1A

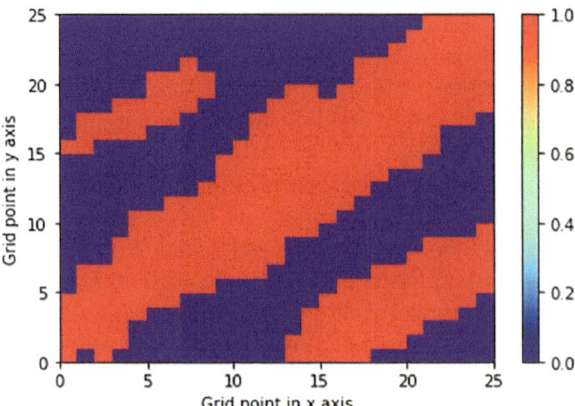

**Fig. 4** Seismic impedance map for Case 1A

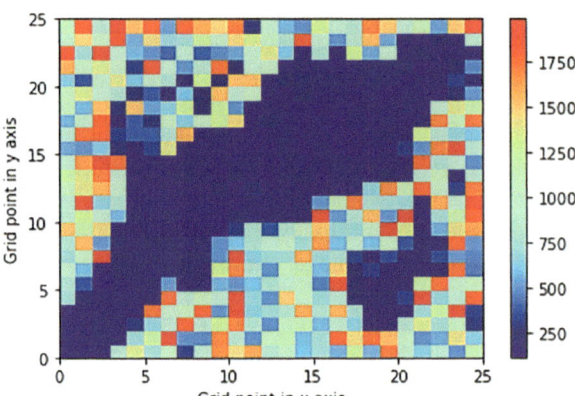

be made geologically realistic, and this process can be further improved by considering expert information regarding the subsurface. This would then enable the use of training images for multi-point simulation using locally varying mean data [45]. As mentioned earlier, the assumptions governing the ensemble generation process can have dramatic effects on the policy that the DRL agent eventually converges to. The DRL agent attempts to identify the trends in channel connectivities in the reservoir models assuming that the channels are the preferred pay facies in the reservoir.

The pixel-wise average of pay-facies across reservoir realizations is shown in Fig. 5. The reservoir DRL agent is trained following the methodology shown in Fig. 1 and is then compared to a single-stage optimization policy. The single-stage policy maximizes the reward over the placement of the 5 wells by placing wells in the grid points with the highest probability of locating a pay-facies region (the same reward function formulation is considered for both single-stage optimization technique and the DRL technique) and is shown in Fig. 6. As evident by this simultaneous well placement policy, general trends in the seismic impedance are used to build a mechanism to exploit the reservoir. In the case where the initial assumptions regarding the reservoir are incorrect (due to imprecision and translation of features), this short-horizon policy would suggest well placements that are not profitable and can skew the expectation of optimality (as evident by the placement of Well 4 in non-pay facies region). In addition, due to the incremental nature of geospatial analysis, future information is not considered for present decision making.

Figure 7 demonstrates the manner in which the reservoir DRL agent is trained to address the problem. The figure shows the increase in the cumulative return with increasing number of episodes the agent is trained. The expected return tapers off

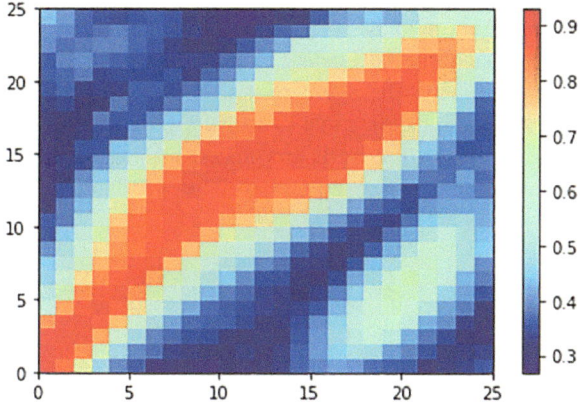

**Fig. 5** Pixelwise average of reservoir facies for the ensemble of models generated using sequential indicator simulation using locally varying mean

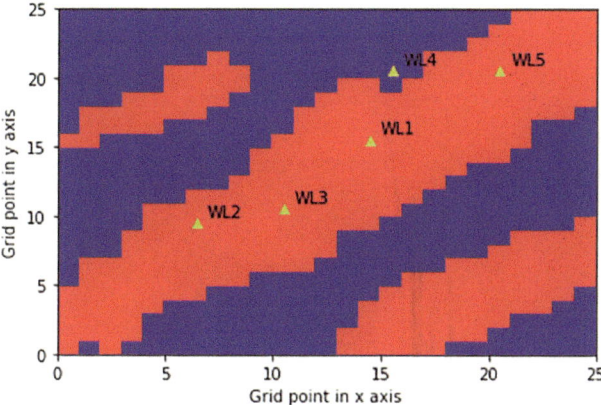

**Fig. 6** Representation of the single-stage optimization policy (following conventional geostatistical well placement methodologies which involve the selection of the most probable well location under well spacing and productivity constraints) on the ground reality

asymptotically and at this stage the policy is assumed to have converged. The asymptotic nature of the convergence also depends on the exploration hyperparameter $\epsilon$ and requires the annealing of the parameter to transition from exploration to exploitation. The trained policy has been demonstrated on the ground reality and several reservoir realizations in Fig. 8 and Fig. 9 respectively. The sequential placement of wells contingent upon prior wells placed can clearly be seen in the figures. Across all reservoir realizations, the first well placed is in the same location. This is because the initial well placement solely depends on the prior seismic data and the state encoding contains no information regarding the reservoir facies. After the first well, the additional information from that well influences the reservoir models and the subsequent well is placed according to the developed model.

In addition to the placement of wells in the true reservoir, it is interesting to visualize the performance of the DRL agent on the ensemble of reservoir realizations (as demonstrated in Fig. 10). The agent is not able to suggest perfect placement of wells across all realizations. In most cases, the DRL agent can generate high rewards. In a few realizations, the wells are placed sub-optimally (for ~12% of the realizations). The placement of subsequent wells depends upon the results of the

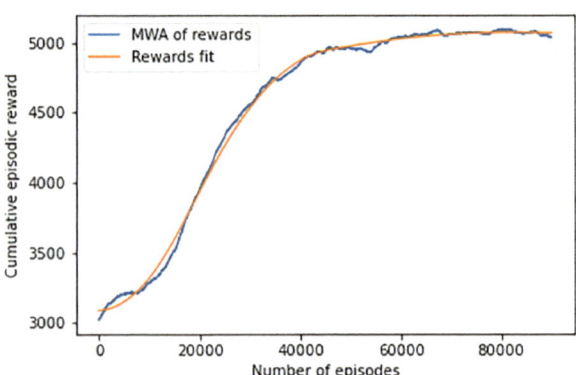

**Fig. 7** Moving average cumulative reward per episode for Case 1A

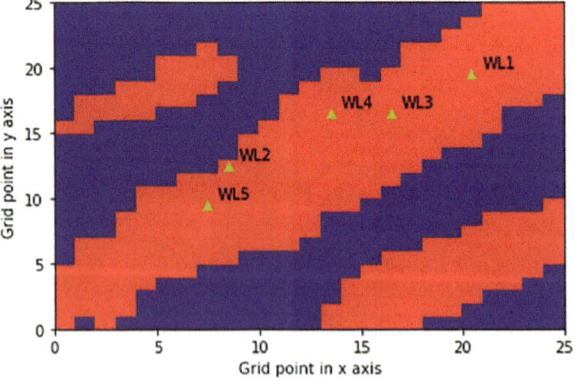

**Fig. 8** Well placement based upon policy generated by DRL agent

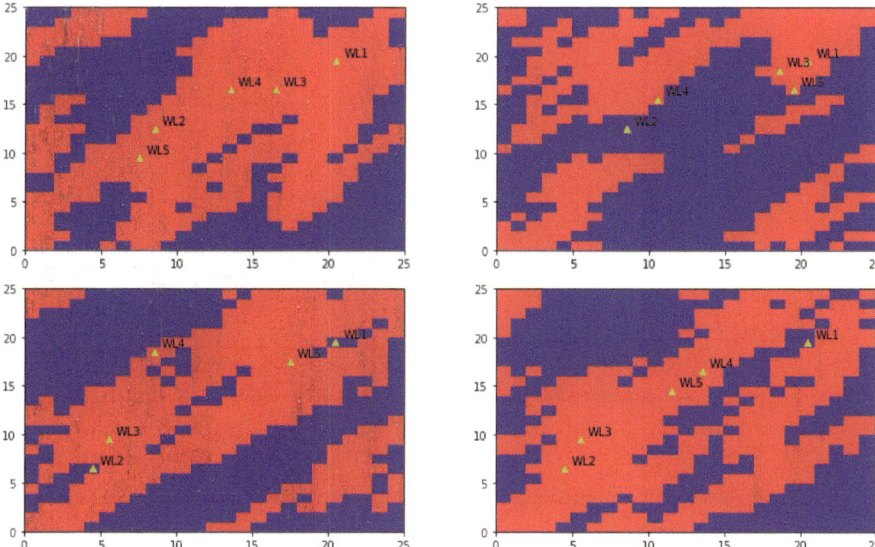

**Fig. 9** Policy demonstrated on selection of realizations from the ensemble generated using indicator simulation

reservoir property analysis at the given well location. For a 2D binary facies case with the objective of placing 5 wells, there are at most 16 facies combinations at the well locations $\left( n_{facies}^{n_{wells}-1} \right)$ generated and the developed policy must account for the same. The process for determination of the optimal policy is similar to pruning the leaf nodes in a decision tree. Figure 11 demonstrates the process for developing the policy and the well configuration developed for the ground truth case has been highlighted. As the major channel is aligned at 45° with respect to the horizontal, successful wells are placed with coordinate where $x \approx y$. When the DRL agent initially fails to place wells in pay regions, it moves towards exploring regions where $x$ deviates from $y$ (the downward facing branches of the tree).

Parameters that govern the performance of the deep Q-learning process such as the learning rate ($\alpha$), mini-batch size, replay buffer size and the exploration parameter can alter the learning rate and eventual policy convergence. The effects of the hyperparameters on the developed policy, assuming the same neural network model configuration and computational resources (2.5 GHz Intel Xeon Processor, 2 Nvidia Tesla K80 computing modules, FDR InfiniBand, 10 Gbps Ethernet), is shown in Table 2. With increase in batch size, there is a decrease in the number of episodes required to convergence and an overall improvement in the final converged policy. The time taken for the convergence of the policy scales nearly hyperbolically with respect to $\tau$, which is the number of episodes between updates to the target network under a DDQN approach, with no dramatic reduction in the reward at convergence of the policy.

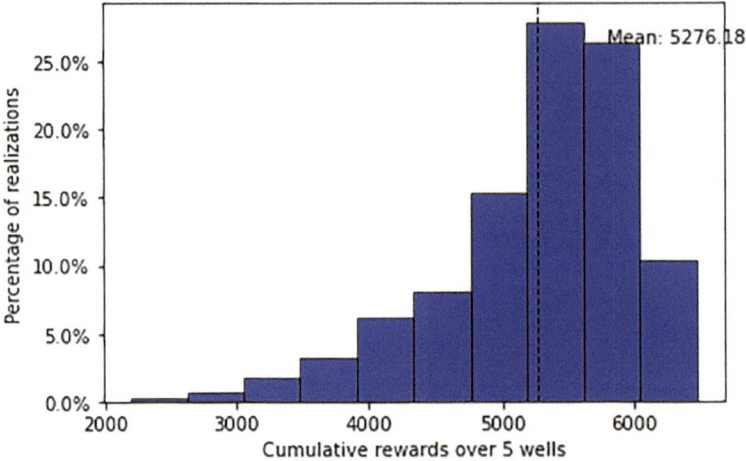

**Fig. 10** Histogram of cumulative rewards from the developed DRL policy, tested on the ensemble of reservoir realizations (for Case 1A)

Case 1B aims to demonstrate the dependence of the developed policy and the training of the agent on the initial seismic information and modeling assumptions. A diffused seismic impedance map, generated via dilation of the seismic impedance map for Case 1A, is now considered leading to a greater uncertainty in the developed ensemble of reservoir realizations. The seismic impedance map and the pixel-wise average of pay-facies is shown in Fig. 12. With the ground truth the same as in Case 1A, the pay facies probability map shows an increase in the uncertainty of pay facies probability at all locations (demonstrated by probability values being closer to 0.5). The reservoir DRL agent is trained on the ensemble of realizations developed using the diffused seismic impedance map and is then tested on the ground model.

The histogram of the reward distribution for this case is shown in Fig. 13. The result shows that the new model performs worse than the other ensemble of reservoir realizations generated using the more precise seismic map. As the DRL agent bases the policy on reservoir realizations generated by modelers and attempts to mimic human decision-making, with the increase in uncertainty regarding the position of channels there is a reduction in the quality of policy developed.

Another significant difference is the computational speed that results from considering more precise information. There was a 19.8% increase in computational time required for convergence of the policy when considering the ensemble of models generated conditioned to less precise seismic information. This is because the DRL trains on markedly different realizations from one case to the next (high uncertainty in the pay facies probability). Figure 14 shows the developed policy tested on the ground truth. The wells are placed further apart as they were trained considering realizations in which the pay facies are aligned roughly in the 45° azimuth direction. However, there is no guarantee that the wells are placed in regions of high channel

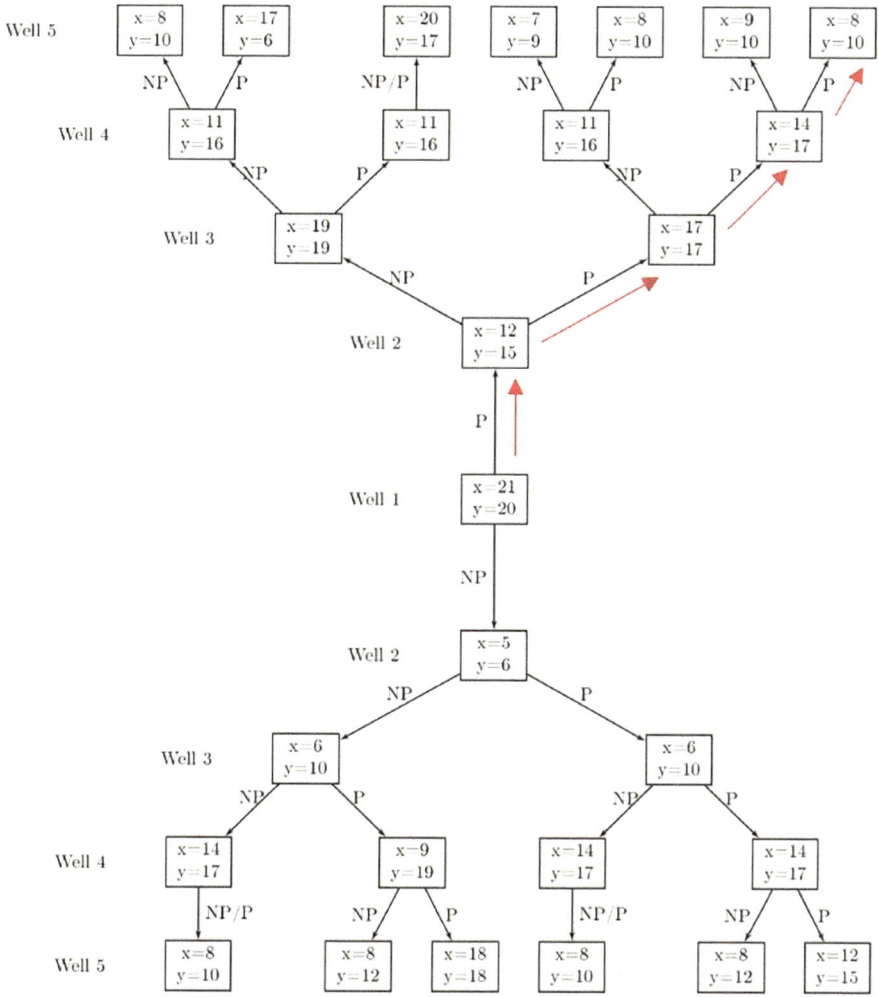

**Fig. 11** Policy for exploitation of the reservoir (NP representing that the placed well is a non-pay facies region and P represents pay facies region). The coordinates of the next well have been shown within the boxes

connectivity in individual realizations due to the high uncertainty associated with the realizations.

Cases 1C and 1C_alternate considers the robustness of the training to faulty interpretations of channel orientation from seismic data. The goal in this case is to demonstrate the variation in the generated policy with increasing uncertainty associated with detected seismic features. The seismic impedance map from Case 1A is considered along with two other seismic impedance maps (shown in Fig. 15). These maps contain the main channel oriented at 35° and 15° with respect to the x-axis (the original seismic has the main channel oriented at 45° with respect to the x-axis)

**Table 2** Effect of hyperparameters on the convergence of the final policy in terms of the quality of the developed policy (demonstrated using the converged cumulative reward) and the computational time

| Batch size | $\tau$ | Convergence time (s) | Reward at convergence |
|---|---|---|---|
| 64 | 1 | 2157.5 | 4873.5 |
| | 4 | 677.3 | 4868.9 |
| | 16 | 308.4 | 4430.8 |
| 128 | 1 | 2211.8 | 4953.4 |
| | 4 | 684.7 | 4887.0 |
| | 16 | 309.4 | 4681.5 |
| 256 | 1 | 2246.0 | 5011.1 |
| | 4 | 683.1 | 4977.2 |
| | 16 | 307.9 | 4906.4 |

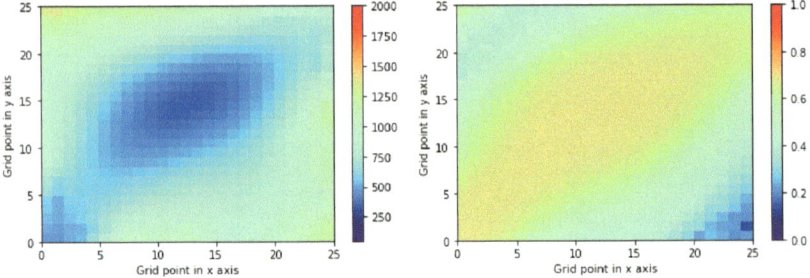

**Fig. 12** Seismic impedance map for Case 1B showing diffused channel case and the pixel-wise average of pay-facies across the ensemble of reservoir realizations. Consider this against the crisper channel description in Case 1A

**Fig. 13** Histogram of cumulative rewards from the developed deep DRL policy trained using the ensemble of reservoir realizations generated using the diffused channel seismic impedance map, tested on the ensemble of reservoir realizations (for Case 1B)

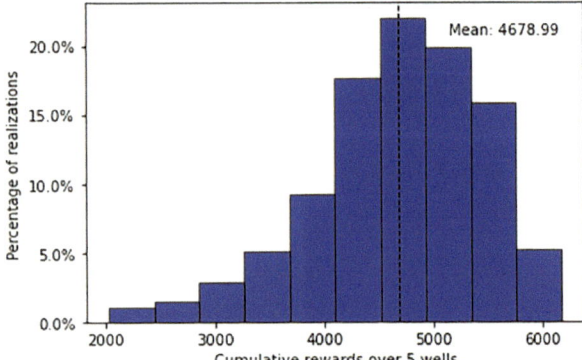

**Fig. 14** Well placement on ground truth suggested by DRL agent trained using the ensemble of realizations generated conditioned to the diffused seismic data

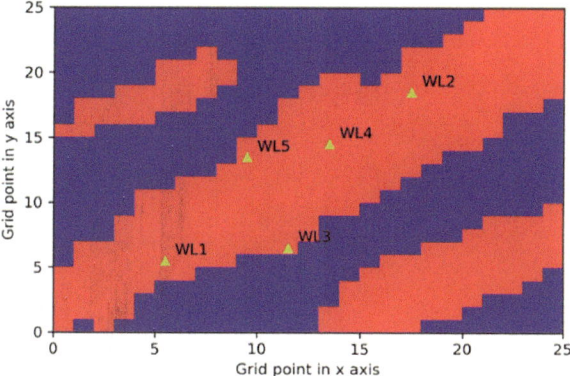

and will be used to generate *two distinct sub-cases* labeled 1C for the 35° seismic and 1C_alternate for the 15° seismic. The seismic impedance maps for Cases 1C and 1C_alternate were generated by rotating individual sections of the seismic impedance map for Case 1A by the prescribed angle followed by diffusion to blend in the gaps. The pixel-wise average of the pay facies across the ensemble of generated realizations have been shown in Fig. 16. By training the DRL agent on these variations of pay facies map, the effect of initial assumptions on the developed policy is demonstrated. In addition to demonstrating the developed policy on the ground truth model (shown in Fig. 17), the histogram of episode rewards for the application of the DRL agent on the realizations developed using the seismic impedance maps has been shown in Fig. 18. As the channel orientation starts deviating significantly from the orientation in the "ground truth" model (from 10° deviation in Case 1C to 30° deviation in Case 1C_alternate), the developed policy performs worse. The wells may end up getting placed in the non-pay regions. In both the sub-cases the reservoir DRL agent attempts to place the wells in the orientation of the major channel in their trained ensemble. Though the policy developed for Case 1C enables the agent to place wells in pay facies regions, these wells are placed in the periphery of the major channel. In Case 1C_alternate, the first well is placed in a non-pay facies location followed by 3 wells placed in pay facies regions. This leads the agent to falsely 'believe' that Well 1 was an anomaly and the major channel may yet be found at 15° orientation with respect to the x-axis. This is ultimately proven false with the 5th and final well. Hence, the initial assumptions upon which the ensemble of reservoir realizations are built can have a dramatic effect on the eventual policy developed with the associated error increasing significantly if models are conditioned to incorrectly interpreted information. This is crucial to demonstrate that the DRL agent attempts to mimic human decision-making and falls to the same pitfalls as decision-makers would if poor reservoir models are used to guide the decision-making.

Case 2 considers the placement of vertical wells in a 3D reservoir ($100 \times 120 \times 10$). The 3D case considers the Stanford V reservoir with the goal of placing 5 vertical wells in an uncertain environment. To extend the 2D case to a 3D one, a modification is made to the state vector representation. The facies at the well location can represented

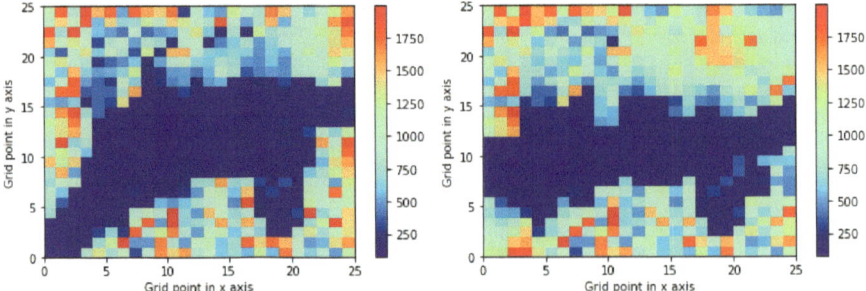

**Fig. 15** Seismic impedance maps for Cases 1C on the left and 1C_alternate on the right considering imprecise seismic data that incorrectly identifies the channel angle

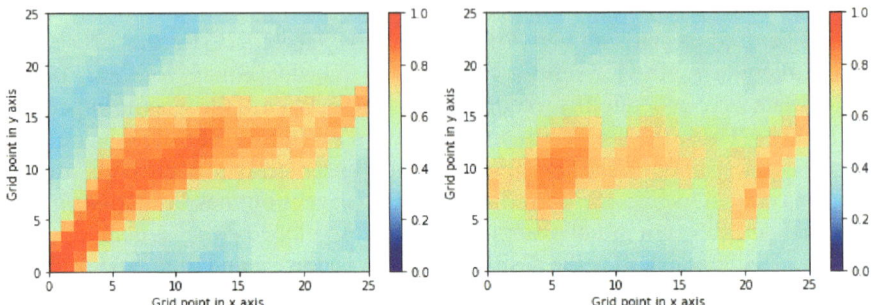

**Fig. 16** Pixel-wise average of pay facies across the ensemble of reservoir realizations generated using seismic data that do not reflect the correct channel orientation for Case 1C (left) and Case 1C_alternate (right)

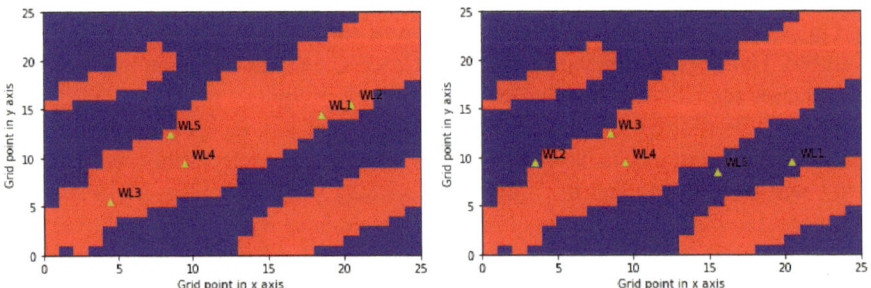

**Fig. 17** Developed policy demonstrated on the ground truth for Cases 1C and 1C_alternate respectively

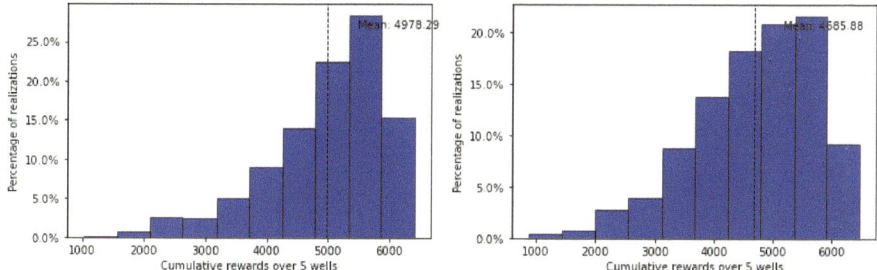

**Fig. 18** Histogram of cumulative rewards from the developed DRL policy trained using the ensemble of reservoir realizations generated using the rotated channel seismic impedance map, tested on the ensemble of reservoir realizations generated using the original seismic data for Cases 1C and 1C_alternate, respectively

as an aggregate of facies encoding at all grid blocks contacted by the vertical well. While treating the 3D model as a stack of 2D models require less computational resources for the determination of the optimal policy, that assumption may result in the failure to capture the facies spatial correlations between different z-slices. The 3D formulation addresses this deficiency albeit at the expense of additional computational resources.

The ground truth and the pixel-wise ensemble of reservoir realization are shown in Figs. 19 and 20. Using the process workflow shown in Fig. 1, the DRL agent efforts to maximize the cumulative expected reward. The 3D formulation converges to a better policy than the case with multiple stacked 2D layers. This is due to the better representation of spatial correlation between reservoir layers. It is clear from the ensemble average map in Fig. 20 that the stochastic realizations conditioned only to the seismic impedance information do not depict the channel trends in the ground truth model accurately. The major channel features are displaced to the northern part of reservoir in Fig. 20 and several key high productivity regions are also consistently missing in the generated ensemble. Though in real-world cases experts add information to the ensemble generation process in the form of prior geologic constraints, the current work has deliberately not attempted to do so, to not introduce bias in the generated ensemble. Constraints such as well productivity, reservoir boundary and channel connectivity have been considered in this formulation.

Figures 21 and 22 demonstrate the well locations corresponding to the optimum DRL policy and single-stage optimization policy respectively on the ground truth. Due to the uncertainty represented in the ensemble, the second and third wells in the optimum DRL policy are placed in an unproductive region (in high probability pay-facies region in the ensemble as seen in Fig. 20). The well placement policy generated using DRL shows a gradual update to the beliefs regarding the pay facies location. The DRL algorithm learns about the non-existence of high productivity regions at the location of wells 2 and 3 and applies that knowledge to the placement of wells 4 and 5. These wells are located well within channel regions. In the single-stage optimization policy (implementing the gradient-based optimization considering the same reward

**Fig. 19** Pixel-wise addition of the ground truth for Case 2

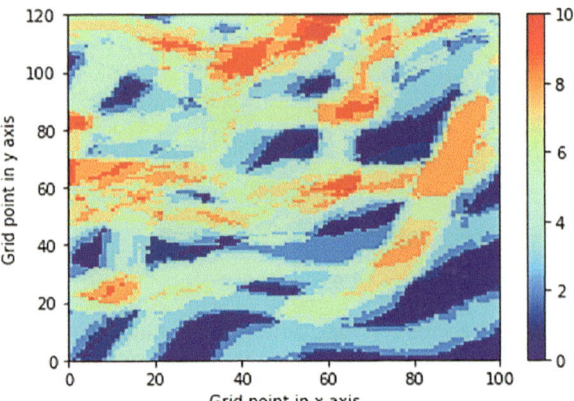

**Fig. 20** Pixel-wise average of pay facies (standardized by dividing the number of layers) across layers averaged over the ensemble of reservoir realizations generated for Case 2

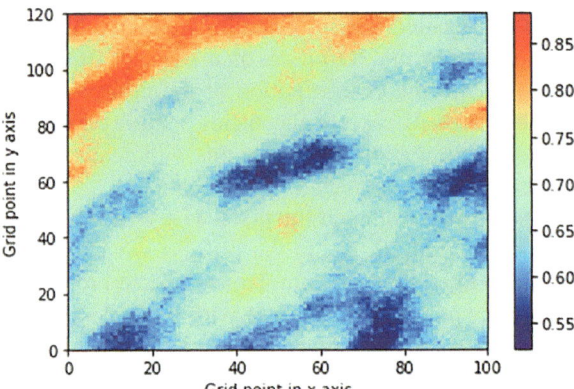

formulation as the DDQN method), every single well placement has an exaggerated effect on the beliefs regarding the location of channels and well placement oscillates wildly from one region of the reservoir to the next. In real-world scenarios, field development decisions may involve the initial drilling of an exploratory well and then, depending on the observation of reservoir properties at the well location, decision would be taken to either step out and drill an offset well near the first well or, in case the first well happens to be dry, to explore a different region of the reservoir. While the single-stage optimization policy and the DRL policy derive well placement strategies with similar overall reward, DRL policy demonstrates a policy that mimics this human decision-making process. As mentioned in Sect. 2, RL techniques are guaranteed to deliver the optimal well placement policy and outperforms techniques that do not account for future actions when applied to multi-stage well location optimization problems only when the value functions for all state-action pairs are accurately quantified. The similarity of performance between the single-stage and DRL policies can be attributed to the early termination and incomplete exploration of the state-action space by the DRL agent. Future research will attempt to quantify the variation in the total reward with increasing well placement decision-stages. This will aid in the quantification of the advantages/disadvantages of human-mimicry in well placement policies. The efficacy of the DRL model has been also demonstrated in another 3-D case study which can be found in [44].

To extend the formulation to continuous state properties, such as permeability, porosity etc., and continuous action properties, such as well operating conditions, greater granularity in well location, optimization of horizontal well trajectory etc., the existing reservoir formulation can be modified to include such features into the reward function. Proxy model to evaluate such actions can also be developed. Another avenue forward is to utilize policy gradient methods [46] to directly estimate the optimal policy without evaluating the value functions. These extensions would speed up the DRL process albeit at the cost of some loss in optimality of the developed policy. These improvements will be the areas of study for future research.

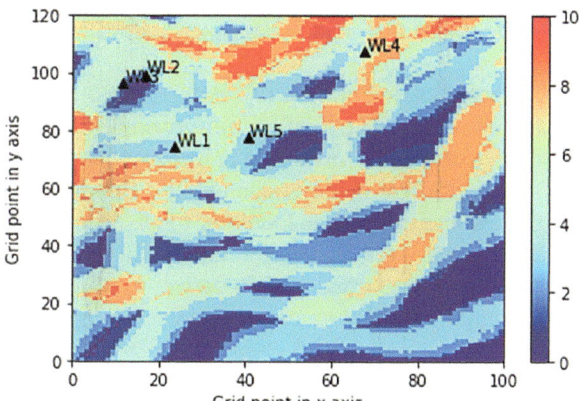

**Fig. 21** Demonstration of the developed DRL policy on the ground truth in the 3-D reservoir case

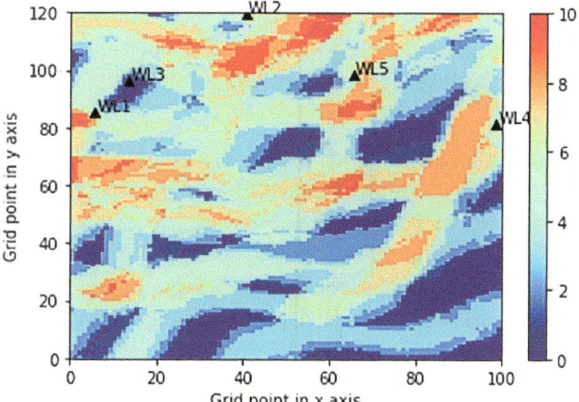

**Fig. 22** Demonstration of the single-stage optimization policy on the ground truth in the 3-D reservoir case

# 5 Discussion

Through the case studies presented in this paper, the authors aim to demonstrate the applicability of reinforcement learning for well location optimization. The paper also presents the pitfalls of poor modeling assumptions that result in sub-optimal well placements. It is evident that, through intelligent sampling, DRL agent can identify high-productivity reservoir regions accounting for constraints in well placement. The effectiveness of reinforcement learning as compared to other well location optimization techniques include,

1. The ability to consider expert information and ease of integration with existing procedures for well location optimization.
2. Taking a long-term view to the well placement problem. Algorithms for well optimization studied in literature take a short horizon view to tackling the well problem, not accounting for the information gained from placing a well. Doing so, in the traditional workflows would require retraining after every well placement action.
3. The ease of integration of optimization methods with existing deep learning infrastructure. The field of reinforcement learning is one of the biggest beneficiaries of the improvements in research into deep learning.
4. Policy gradient methods can aid in the consideration of continuous state-action spaces leading to consideration of diverse set of reservoir properties. Though this comes at the expense of the loss of guarantee of an optimal solution, policy gradient approximations can deliver excellent solutions to the problem at hand if computational resources are scarce.

The authors plan to study policy gradient methods and extend the current formulation to account for the placement of horizontal wells. A comparison between the results developed using temporal difference (TD) methods (as shown in the current work) and policy gradient methods can lay the path for further studies into the use of reinforcement learning for well location optimization.

Reinforcement learning is not a panacea and would not solve all issues associated with the well placement problem. Some of the limitations of reinforcement learning include,

1. Its inherent dependence on the assumptions behind the modeling of the environment (which dictates the transition probabilities, $T(s'|s, a)$, leading to the development of sub-optimal policies if based on unrealistic modeling assumptions.
2. Its high computational expense. It is suitable for multi-step optimization problems but not recommended for single-stage optimization problems.
3. Its need for proxies for the reward to speed up evaluation of candidate wells (or actions in a particular state). This is because full reservoir simulation for each well placement decision can be computationally expensive.
4. Its low interpretability due to the utilization of deep learning models as functional approximators.

# 6   Conclusion

Reinforcement learning is useful to find the optimal solutions of multi-stage decision problems, especially those for which there are feedforward effects of decisions i.e., future decisions are affected by decisions taken in past stages. Due to the computational intensity of the application of RL, it may not be useful for single-stage optimization problems or for problems with extremely low uncertainty regarding the state-action space.

The problem of placement of wells in an uncertain environment can be formulated as a Markov decision process. The authors have demonstrated a novel approximate dynamic programming framework for addressing the problem. Reinforcement learning provides a unique framework for automating decision-making by considering several scenarios and extending the optimal solution to the sub-problems (i.e., location of each well) to generate a comprehensive solution to exploit the reservoir. Also, due to the ease of integration of expert information and reutilization of existing tools, the policies developed using reinforcement learning provide a geostatistically and petrophysically consistent framework for addressing the problem of optimization of well location. By the addition of intelligent sampling techniques and the use of approximately greedy methods for policy determination, the process of locating wells for exploiting reservoir resources can be sped up. The work also utilizes proxy models developed using regression and deep learning models that allow for the faster evolution of optimum well locations. The selection of the reward function dictates the convergence of the DRL algorithms and poor reward formulation may lead to the development of non-optimal policies to solve the optimization problem. The paper also presents cases that demonstrate that the policies developed using reinforcement learning are superior to existing single-stage optimization techniques, but the quality of solutions developed is dependent on the accuracy and precision of the prior models in the ensemble and on the parameters that drive the DRL process.

# Appendix

## *Neural Network Architecture for Different Case Studies*

For Case 1, the multi-perceptron consists of two hidden dense layers with 1000 nodes each (bias excluded) and uniform Glorot uniform initializer. The activation for both the hidden layers was rectified linear units (ReLU). The output layer had a linear activation with the number of nodes equaling the size of the action-space i.e., 625.

For Case 2, the multi-layer perceptron consisted of three hidden layers with 1000, 3000 and 6000 nodes respectively and an output layer with 12000 nodes with Glorot uniform activation. Activations in hidden layers were ReLU and the output layer had linear activation and no bias.

For all cases, the layers are fully connected. The memory buffer size is 1000 for the first case while for Case 2, the memory buffer size considered is 10,000. The hyperparameters $\alpha$, the learning rate, and $\tau$, interval between updates, is set to $5*10^{-4}$ and 4 respectively. Hyperparameter tuning was conducted to determine the optimum values of $\alpha$ ($[5*10^{-4}, 5*10^{-3}, 0.05, 0.1]$) and $\tau$ ($[1, 4, 16]$); the respective lists show the discrete values considered during hyper-parameter tuning. The selection of the values for the hyperparameters was considered by trading off computational time with the optimality of the developed policy. The exploration component $\epsilon$ is annealed according to the following equation.

$$\epsilon_N = \frac{1}{N}$$

where $N$ is the episode number.

## *Visualization of Convergence*

See Figs. 23, 24, 25, 26 and 27.

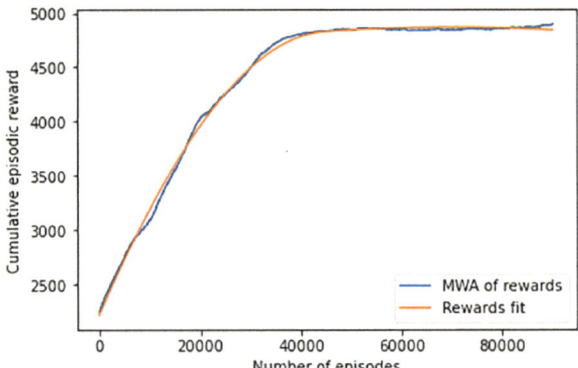

**Fig. 23** Learning trend for DRL agent trained on ensemble of reservoir realizations generated from the diffused channel seismic map (Case 1B)

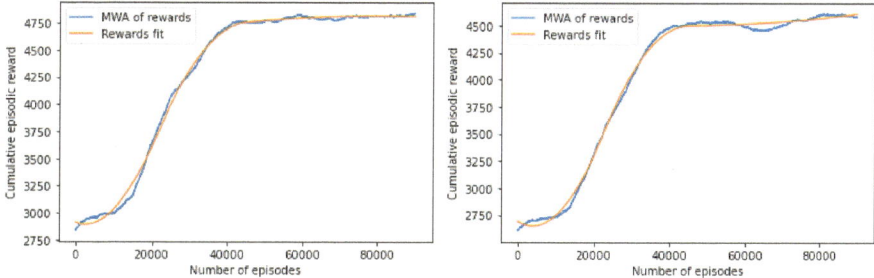

**Fig. 24** Learning trend for DRL agent trained on ensemble of reservoir realizations generated from the rotated channel seismic map for Case 1C (left) and 1C_alternate (right)

**Fig. 25** Convergence of the final policy for Case 2

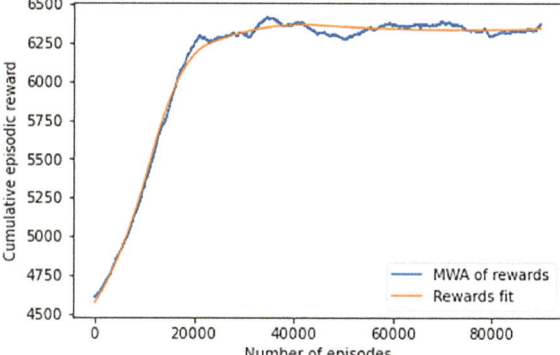

**Fig. 26** Histogram of rewards for testing the developed policy on the ensemble of reservoir realizations for Case 2

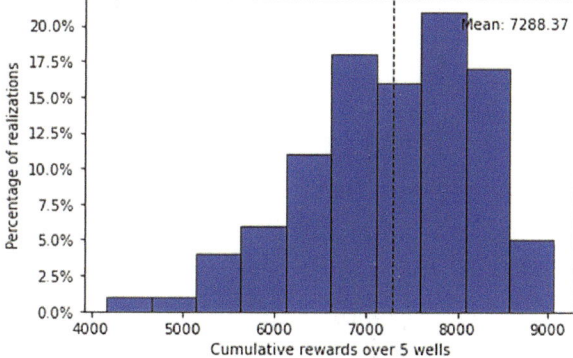

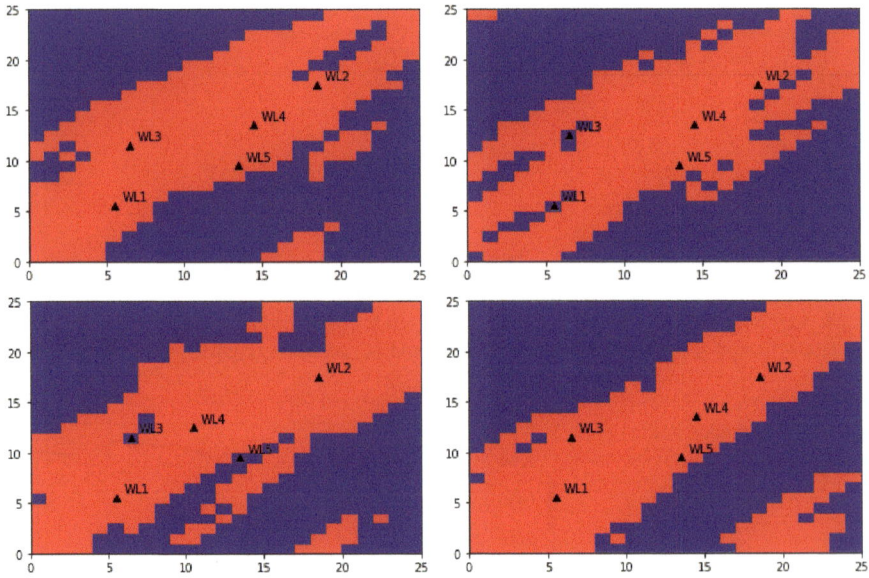

**Fig. 27**  Policy demonstrated on selection of realizations from the ensemble for Case 1B

# References

1. Devine, M., Lesso, W.: Models for the minimum cost development of offshore oil fields. Manag. Sci. **18**(8), B378–B387 (1972)
2. Rosenwald, G.W., Green, D.W.: A method for determining the optimum location of wells in a reservoir using mixed-integer programming. SPE J. **14**(01), 44–54 (1974)
3. van den Heever, S.A., Grossmann, I.E.: An iterative aggregation/disaggregation approach for the solution of a mixed-integer nonlinear oilfield infrastructure planning model. Ind. Eng. Chem. Res. **39**(6), 1955–1971 (2000)
4. Sarma, P., Chen, W.H.: Efficient well placement optimization with gradient-based algorithms and adjoint models. In: Intelligent Energy Conference and Exhibition, Amsterdam, The Netherlands (2008)
5. Bittencourt, A.C., Horne, R.N.: Reservoir development and design optimization. In: SPE Annual Technical Conference and Exhibition, San Antonio, Texas, USA (1997)
6. Glover, F.: Tabu search—Part I. ORSA J. Comput. **1**(3), 190–206 (1989)
7. Glover, F.: Tabu search—Part II. ORSA J. Comput. **2**(1), 4–32 (1990)
8. Güyagüler, B., Horne, R.N.: Uncertainty assessment of well-placement optimization. SPE Reserv. Eval. Eng. **7**(01), 24–32 (2004)
9. Ozdogan, U., Horne, R.N.: Optimization of well placement under time-dependent uncertainty. SPE Reserv. Eval. Eng. 135–145 (2006)
10. Montes, G., Bartolome, P., Udias, A.L.: Use of genetic algorithms in well placement optimization. In: SPE Latin American and Caribbean Petroleum Engineering Conference, Buenos Aires, Argentina (2001)
11. Morales, A.N., Nasrabadi, H., Zhu, D.: Modified genetic algorithm for horizontal well placement optimization in gas condensate reservoirs. In: SPE Annual Technical Conference and Exhibition, Florence, Italy (2010)

12. Abdul-Latif, B.L., Edem, T.D., Hikmahtiar, S.: Well placement optimisation in gas-condensate reservoirs using genetic algorithms. In: SPE/IATMI Asia Pacific Oil & Gas Conference and Exhibition, Jakarta, Indonesia (2017)
13. Narayanasamy, R., Davies, D.R., Somerville, J.M.: Well location selection from a static model and multiple realisations of a geomodel using productivity potential map technique. In: SPE Europec/EAGE Annual Conference and Exhibition, Vienna, Austria (2006)
14. Lee, K., Lim, J., Yoon, D., Jung, H.: Prediction of shale-gas production at Duvernay formation using deep-learning algorithm. SPE J. **24**(06), 2423–2437 (2019)
15. Jin, Y., Shen, Q., Wu, X., Chen, J., Huang, Y.: A physics-driven deep-learning network for solving nonlinear inverse problems. Petrophysics **61**(01), 86–98 (2020)
16. Alpak, F.O., Araya-Polo, M., Onyeagoro, K.: Simplified dynamic modeling of faulted turbidite reservoirs: a deep-learning approach to recovery-factor forecasting for exploration. SPE Reserv. Eval. Eng. **22**(04), 1240–1255 (2019)
17. Pan, S.J., Yang, Q.: A survey on transfer learning. IEEE Trans. Knowl. Data Eng. **22**(10), 1345–1359 (2010)
18. Levinson, R., Weber, R.: Chess neighborhoods, function combination, and reinforcement learning. Comput. Games **2063**, 133–150 (2000)
19. Tedrake, R., Zhang, T.W., Seung, H.S.: Stochastic Policy Gradient Reinforcement Learning on a Simple 3D Biped. Sendai (2004)
20. Lopez-Martin, M., Carro, B., Sanchez-Esguevillas, A.: Application of deep reinforcement learning to intrusion detection for supervised problems. Expert Syst. Appl. **141** (2020)
21. Soranzo, E., Guardiani, C., Saif, A., Wu, W.: A reinforcement learning approach to the location of the non-circular critical slip surface of slopes. Comput. Geosci. **166**(0098–3004), 105182 (2022)
22. Luo, F., Feng, B., Wang, H.: Automatic first-arrival picking method via intelligent Markov optimal decision processes. J. Geophys. Eng. **18**, 406–417 (2021)
23. Schaul, T., Quan, J., Antonoglou, I., Silver, D.: Prioritized experience replay. Comput. Res. Repos. (CoRR), vol. abs/1511.05952, 2016.
24. Ipek, E., Mutlu, O., Martinez, J.F., Caruana, R.: Self-optimizing memory controllers: a reinforcement learning approach. SIGARCH Comput. Archit. News **36**(3), 39–50 (2008)
25. van Hasselt, H., Guez, A., Silver, D.: Deep Reinforcement Learning with Double Q-learning (2015)
26. Wang, Z., Schaul, T., Hessel, M., van Hasselt, H., Lanctot, M., Freitas, N.: Dueling network architectures for deep reinforcement learning. In: International Conference on Machine Learning (2016)
27. Bellman, R.: A Markovian decision process. J. Math. Mech. **6**(5), 679–684 (1957)
28. Bellman, R.: Dynamic Programming. Dover Publications, Mineola, NY (2003)
29. Tesauro, G.: TD-Gammon, a self-teaching backgammon program, achieves master-level play. Neural Comput. **6**(2), 215–219 (1994)
30. Mnih, V., Kavukcuoglu, K., Silver, D., Graves, A., Antonoglou, I., Wierstra, D., Riedmiller, M.: Playing Atari with Deep Reinforcement Learning (2013)
31. Cotter, A., Shamir, O., Srebro, N., Sridharan, K.: Better mini-batch algorithms via accelerated gradient methods. In: Advances in Neural Information Processing Systems (2011)
32. Ma, Y.Z., Gomez, E., Seto, A.: Coupling spatial and frequency uncertainty analyses in reservoir modeling: example of Judy Creek Reef complex in Swan Hills, Alberta, Canada. AAPG Mem. **96**, 159–173 (2011)
33. Doyen, P., Psaila, D., Strandenes, S.: Bayesian sequential indicator simulation of channel sands from 3-D seismic data in the Oseberg field, Norwegian North Sea. In: SPE Annual Technical Conference and Exhibition, New Orleans, Louisiana (1994)
34. Qi, L., Carr, T., Goldstein, R.H.: Geostatistical three-dimensional modeling of oolite shoals, St. Louis Limestone, southwest Kansas. Am. Asso. Petrol. Geol. Bull. **91**(1), 69–96 (2007)
35. Rahimpour-Bonab, H., Aliakbardoust, E.: Pore facies analysis: incorporation of rock properties into pore geometry based classes in a Permo-Triassic carbonate reservoir in the Persian Gulf. J. Geophys. Eng. **11**(3), 035008 (2014)

36. Oraki Kohshour, I., Ahmadi, M., Hanks, C.: Integrated geologic modeling and reservoir simulation of Umiat: a frozen shallow oil accumulation in national petroleum reserve of Alaska. J. Unconv. Oil Gas Resour. **6**(2213–3976), 4–27 (2014)
37. Li, L., Qu, J., Wei, J., Xia, F., Gao, J., Liu, C.: Facies-controlled geostatistical porosity model for estimation of the groundwater potential area in Hongliu Coalmine, Ordos Basin, China. ACS Omega **6**(15), 10013–10029 (2021)
38. Chehrazi, A., Rezaee, R.: A systematic method for permeability prediction, a Petro-Facies approach. J. Petrol. Sci. Eng. **82–83**(0920–4105), 1–16 (2012)
39. Zeiler, M.D., Krishnan, D., Taylor, G.W., Fergus, R.: Deconvolutional networks. In: 2010 IEEE Computer Society Conference on Computer Vision and Pattern Recognition (2010)
40. Men, K., Chen, X., Zhang, Y., Zhang, T., Dai, J., Yi, J., Li, Y.: Deep deconvolutional neural network for target segmentation of nasopharyngeal cancer in planning computed tomography images. Front. Oncol. **7**, 315 (2017)
41. Vera-Olmos, F.J., Malpica, N.: Deconvolutional Neural Network for Pupil Detection in Real-World Environments. Cham (2017)
42. Cesa-Bianchi, N., Gentile, C., Lugosi, G., Neu, G.: Boltzmann exploration done right. In: Advances in Neural Information Processing Systems (2017)
43. Asadi, K., Littman, M.L.: An alternative softmax operator for reinforcement learning. In: Proceedings of the 34th International Conference on Machine Learning (2017)
44. Dawar, K.: Reinforcement Learning for Well Location Optimization. The Pennsylvania State University, State College, Pennsylvania (2021)
45. Strebelle, S., Payrazyan, K., Caers, J.: Modeling of a deepwater turbidite reservoir conditional to seismic data using multiple-point geostatistics. In: SPE Annual Technical Conference and Exhibition, San Antonio, Texas (2002)
46. Sutton, R.S., Barto, A.G.: Reinforcement Learning: An Introduction, 2nd edn. MIT Press, Cambridge, Massachusetts (2018)

# Compression-Based Modelling Honouring Facies Connectivity in Diverse Geological Systems

Tom Manzocchi⊙, Deirdre A. Walsh⊙, Javier López-Cabrera, Marcus Carneiro, and Kishan Soni

**Abstract** In object- or pixel-based modelling, facies connectivity is tied to facies proportion as an inevitable consequence of the modelling process. However, natural geological systems (and rule-based models) have a wider range of connectivity behaviour and therefore are ill-served by simple modelling methods in which connectivity is an unconstrained output property rather than a user-defined input property. The compression-based modelling method decouples facies proportions from facies connectivity in the modelling process and allows models to be generated in which both are defined independently. The two-step method exploits the link between the connectivity and net:gross ratio of the conventional (pixel- or object-based) method applied. In Step 1 a model with the correct connectivity but incorrect facies proportions is generated. Step 2 applies a geometrical transform which scales the model to the correct facies proportions while maintaining the connectivity of the original model. The method is described and illustrated using examples representative of a poorly connected deep-water depositional system and a well-connected fluid-driven vein system.

**Keywords** Facies connectivity · Amalgamation ratio · Compression algorithm

## 1 Introduction

The objectives of this contribution are to demonstrate an important limitation in object- and pixel-based facies modelling with respect to facies connectivity, and to highlight the main features of the compression-based modelling method developed to overcome this limitation. Further details of the method can be found elsewhere [1–3].

A modified and extended version of this work has been submitted to the Geostats 2021 Special Issue in the journal Mathematical Geosciences.

T. Manzocchi (✉) · D. A. Walsh · J. López-Cabrera · M. Carneiro · K. Soni
iCRAG (SFI Research Centre in Applied Geosciences) and Fault Analysis Group, UCD School of Earth Sciences, University College Dublin, Dublin 4, Ireland
e-mail: Tom.Manzocchi@ucd.ie

S. A. Avalos Sotomayor et al. (eds.), *Geostatistics Toronto 2021*, Springer Proceedings in Earth and Environmental Sciences, https://doi.org/10.1007/978-3-031-19845-8_8

## 2  Connectivity in Facies Models and Natural Systems

Two different ways of considering facies connectivity are important to this work (Fig. 1). Global connectivity refers to characteristics of the largest connected cluster of objects and have been examined for object-based models (OBM) in the context of percolation theory using the net:gross ratio (NTG). The simplest models consisting of aligned cuboids have a well-defined connectivity threshold at $NTG_C \approx$ 28% [4]. Connectivity thresholds for OBM containing geometrically representative, stationary, three-dimensional systems of more geometrically diverse [5] or geologically realistic [1, 6] elements show similar or lower thresholds (Fig. 2a), with the lowest $NTG_C$ for systems of more anisotropic and misaligned objects. Pixel-based models (PBM) built using the sequential indicator (SIS), or truncated Gaussian (TGS) methods have similar thresholds as OBMs, with $NTG_C \leq 28\%$, as do models built using the pixel-based SNESIM multiple point (MPS) method even if the training images used to create the models have much lower connectivity (Fig. 2b, [7]). The inability of the SNESIM MPS method to honour the connectivity of the training image is seldom acknowledged but is a recognised restriction of the method [3, 8–10].

These consistent thresholds at $NTG_C \leq 28\%$ in OBM or PBM have led some to conclude that they are transportable to natural geological systems [6]. It is hard to test this generalisation directly since it is impossible to estimate global 3D connectivity from limited outcrop data, and very difficult to do so from subsurface data. Therefore, local measures of connectivity such as the amalgamation ratio (AR, Fig. 1b) are useful. AR can be measured in OBM or natural systems but not in PBM. Crossplots of AR versus NTG for numerous natural depositional systems show that AR $\ll$ NTG (Fig. 3a, [1, 11, 12]). This contrasts with object-based models, for which AR = NTG if all objects are of constant thickness, or is slightly lower for variable sized objects ([1, 2, 7], Fig. 3b). Hence, the local connectivity behaviour of OBM and natural systems are not the same, and so it is unlikely that their global connectivity behaviour is.

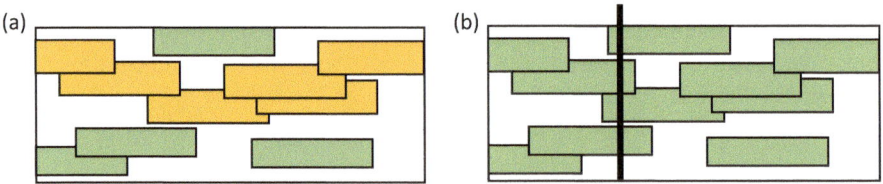

**Fig. 1** Local and global connectivity. **a** Global connectivity is determined as a function of the properties of the largest connected cluster of objects. Here, that cluster occupies ca. 60% of the total volume of objects, and spans entirely across the model. **b** The simplest definition of local connectivity is amalgamation ratio (AR) measured in a 1D vertical sample. In this example, one of the four object bases sampled is amalgamated with a lower object, and hence AR = 0.25

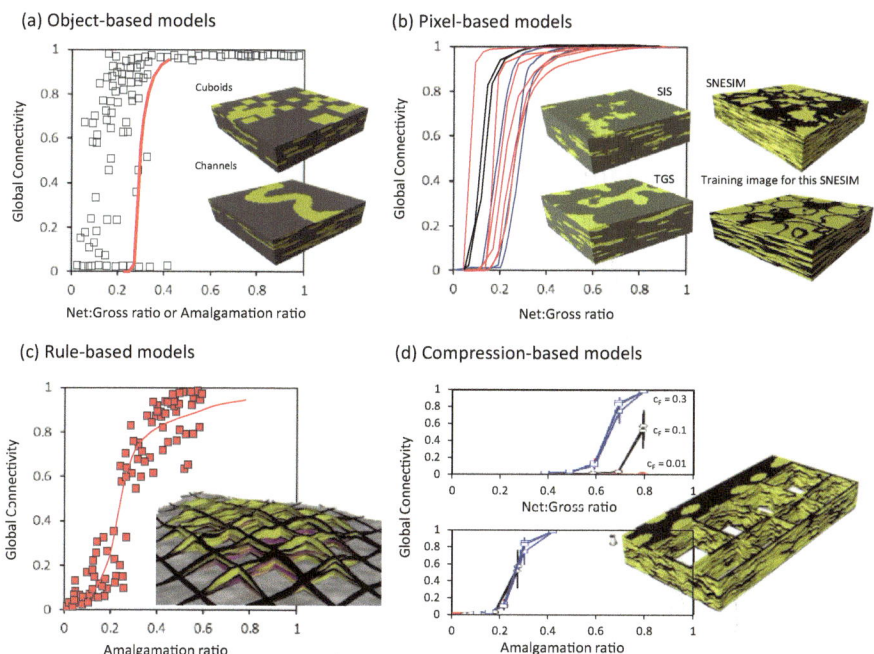

**Fig. 2** Global connectivity as a function of NTG or AR for: **a** object-based models. Red line: trend line for high resolution models of aligned cuboids [4]. Black points: individual sinuous channels models [6]. **b** Pixel-based models generated with different methods and assumptions. Black: SIS. Red: TGS. Blue: SNESIM [7]. **c** Rule-based models. Gross trend and individual points for lobate models generated with different erosion rules. All these models have NTG ≈0.8 [14]. **d** Compression-based models with different compression factors ($c_F$) [1]

Rule-based facies models (RMB) are created by stacking objects in stratigraphic order using geometrical rules that mimic the depositional processes, and are recognised from a qualitative perspective as being more geologically realistic than object- or pixel-based models [13]. Both local and global connectivity can be measured in rule-based models. Different rules governing erosion and aggradation of the depositional elements provide models with a diversity of local connectivity representative of natural systems ([14], Fig. 3c). Like OBMs and PBMs, RBMs have a well-defined global connectivity threshold. However, this occurs at a critical amalgamation ratio ($AR_C \approx 28\%$ [15], Fig. 3c) rather than at a critical NTG, since RBMs can remain at low connectivity to very high NTG. The RBMs examined [14, 15] had approximately circular bodies, but by analogy with OBM, it is likely that $AR_C$ for channelized RBMs will be lower.

Taken together, these observations of global and local connectivity in natural systems and in OBMs, PBMs and RBMs suggest that they all have a connectivity threshold at $AR_C = 28\%$ (flat lying circular elements) or lower (variably oriented elongate elements). In OBM, AR ≈ NTG, so object-based models have $NTG_C \approx AR_C$. Rule-based models have more degrees of freedom than object- or pixel-based

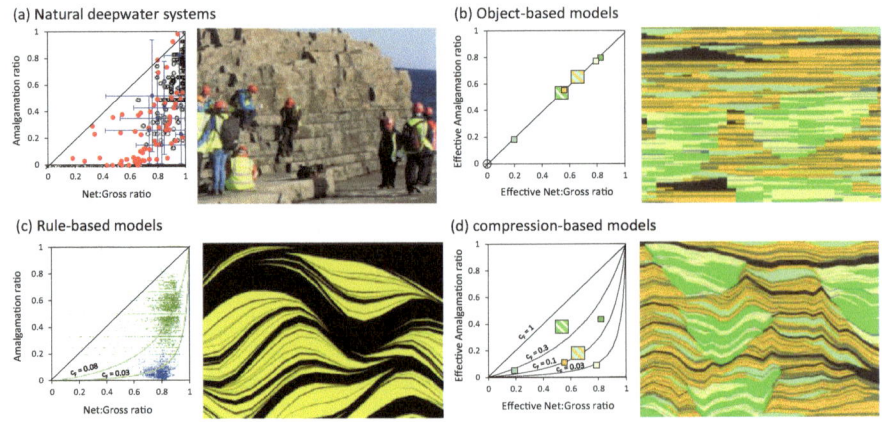

**Fig. 3** Local connectivity (AR) as a function of NTG. **a** Compilations of non-hierarchical (red, [1]) and hierarchical lobe (black, [11]) and channel (blue, [12]) systems. **b** Multi-facies hierarchical OBM. The points show effective values for the 2D cross-sectional model shown, using the same colours as the model. **c** 1D measurements through two different hierarchical rule-based models generated with different erosion rules. The curves show representative model $c_F$ values [15]. **d** Multi-facies hierarchical compression-based OBM. Symbols as **b**. The curves show $c_F$ values representative of the different facies

models and no link between AR and NTG. Therefore, RBM have no intrinsic value of $NTG_C$, and in this respect natural systems are likely to behave similarly to RBM.

## 3    Compression-Based Facies Modelling

The considerations above imply that object- and pixel- methods (including pixel-based MPS) are incapable of creating models with realistically diverse relationships between connectivity and NTG. The compression method was developed to overcome this by providing a means of modifying object-based models so that can have low connectivity at high NTG ([1, 2], Fig. 2d). Compression-based models can be created with user-defined trends of local connectivity representative of natural systems (Fig. 3d). The compression-based geometrical transformation can be applied to pixel-based as well as object-based models, implying that it can be used to create facies models which are both conditioned to well data, and constrained by user-defined facies connectivity [3, 7, 10].

Compression-based facies modelling is a two-step process (Fig. 4, [1, 7]). In Step 1, a conventional object- or pixel-based model is created with a net:gross value equal to the target AR value (Fig. 4a, b). In Step 2 the thickness of cells of the different facies are expanded or compressed vertically by particular factors. This grid transformation modifies the facies proportions but does not alter the grid topology, and therefore the facies connectivity (e.g. AR) is unchanged (Fig. 4a). Compression-based models

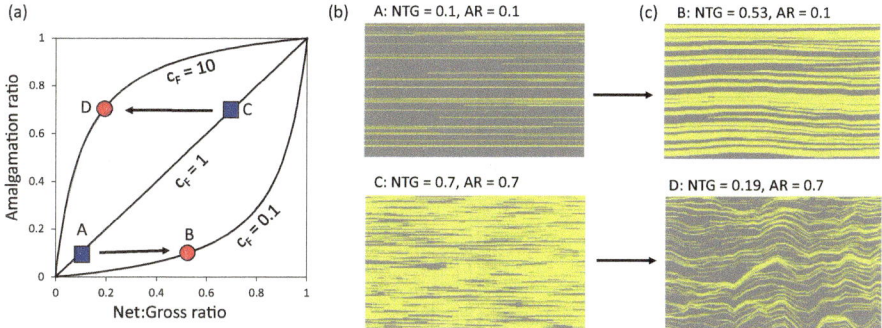

**Fig. 4** The compression algorithm for a two facies OBM (foreground facies yellow, background facies grey). **a** The Step 1 models (blue squares) have AR = NTG ($c_F = 1$). These are transformed to Step 2 models (red circles) with lower ($c_F < 1$, e.g. system B) or higher ($c_F < 1$, e.g. system D) connectivity than the original Step 1 models. Example Step 1 (**b**) and Step 2 (**c**) models

can have higher or lower connectivity than would be present in a conventional model at the same NTG (Fig. 4c). The extent to which connectivity and NTG differ can be expressed by a facies-specific compression factor (cF, Figs. 3d, 4a), which can be estimated by reference to natural system analogues (Fig. 3a).

The two simple systems generated in this example (Fig. 4) are shown alongside photos of the geological systems that inspired them in Fig. 5. Depositional systems (particularly deep marine ones) often consist of laterally extensive sand bodies often entirely enclosed by shale (Fig. 5a, b). They are characterised by AR $\ll$ NTG (Fig. 2a) and are modelled with $c_F < 1$. (Fig. 4a). Other geological systems such as fluid driven injectite or diagenetic vein system can be more connected than a random system at the same NTG value (Fig. 5c, d), and must be modelled with $c_F > 1$ (Fig. 4a).

# 4 Conclusions

The amalgamation ratio in object-based models is an unconstrained output property that is approximately equal to the model net:gross ratio, and object- and pixel-based facies models have connectivity thresholds at $NTG_C = AR_C \leq 28\%$. In natural geological systems and rule-based facies models NTG $\neq$ AR, but $AR_C$ takes similar values and has similar sensitivities as it does in object-based models. The compression algorithm is a geometrical grid transformation which exploits these relationships to provide object- and pixel-based models with user-defined connectivity that is independent of NTG.

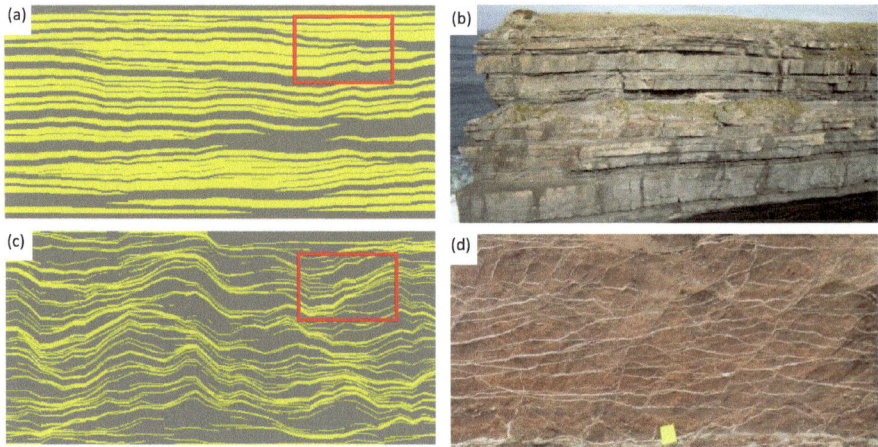

**Fig. 5** 2D cross-sectional compression-based models and the geological systems that inspired them. **a** Model of poorly connected sands (yellow) in shale. **b** Deepwater lobe deposits of the Ross Formation, Loop Head, Ireland. **c** Model of a network of well-connected veins (yellow) in mudstone. **d** Diagenetic gypsum veins in the Mercia Mudstone Formation, Watchet, England. The red rectangles in **a**, **c** are representative of the system sizes photographed in **b**, **d**

# References

1. Manzocchi, T., Walsh, J.J., Tomasso, M., Strand, J., Childs, C., Haughton, P.D.W. Static and dynamic connectivity in bed-scale models of faulted and unfaulted turbidites. In: Jolley, S.J., Barr, D., Walsh, J.J., Knipe, R.J. (eds.) Structurally Complex Reservoirs. Geological Society, London, Special Publications, pp. 309–336 (2007)
2. Manzocchi, T., Zhang, L., Haughton, P.D.W., Pontén, A.: Hierarchical parameterization and compression-based object modelling of high net:gross but poorly amalgamated deep-water lobe deposits. Pet. Geosci. **26**, 545–567 (2020)
3. Walsh, D.A., Manzocchi, T.: A method for generating geomodels conditioned to well data with high net:gross ratios but low connectivity. Mar. Pet. Geol. **129**, 105104 (2021)
4. King, P.R.: The connectivity and conductivity of overlapping sand bodies. In: Buller, A.T., et al. (eds.) North Sea Oil and Gas Reservoirs—II, pp. 353–362. Springer, Netherlands (1990)
5. Garboczi, E.J., Snyder, K.A., Douglas, J.F., Thorpe, M.F.: Geometrical percolation-threshold of overlapping ellipsoids. Phys. Rev. E **52**, 819–828 (1995)
6. Larue, D.K., Hovadik, J.: Connectivity of channelized reservoirs: a modelling approach. Pet. Geosci. **12**, 291–308 (2006)
7. Walsh, D.A., Manzocchi, T.: Connectivity in pixel-based facies models. Math. Geo-sci. **53**, 415–435 (2021)
8. Strebelle, S.: Multiple-point statistics simulation models: pretty pictures or decision-making tools? Math. Geosci. **53**, 267–278 (2021)
9. Tahmasebi, P.: Multiple point statistics: a review. In: Daya Sagar, B.S., Cheng, Q., Agterberg, F. (eds.) Handbook of Mathematical Geosciences. Springer International Publishing, pp. 613–643 (2018)
10. Walsh, D.A., López-Cabrera, J., Manzocchi, T.: The suitability of different training images for producing low connectivity, high net:gross pixel-based MPS models. This volume (2022)
11. Zhang, L., Manzocchi T., Pontén, A.: Hierarchical parameterisation and modelling of deep-water lobes. In: Petroleum Geostatistics 2015, cp-456–00023, EAGE, Biarritz (2015)

12. Soni, K., Manzocchi, T., Haughton, P., Carneiro, M.: Hierarchical characterization and modelling of deep-water slope channel reservoirs. In: Norway Subsurface Conference 2020. Society of Petroleum Engineers, SPE-200763-MS (2020)
13. Pyrcz, M.J., Sech, R.P., Covault, J.A., Willis, B.J., Sylvester, Z., Sun, T.: Stratigraphic rule-based reservoir modelling. Bull. Can. Pet. Geol. **63**, 287–303 (2015)
14. López-Cabrera, J., Manzocchi, T., Haughton, P.D.W.: Rule-based models of deep-water lobes and their influence on connectivity. In: 81st EAGE Conference and Exhibition. EAGE, London, pp. 1–5 (2019)
15. López-Cabrera, J., Manzocchi, T., Haughton, P.D.W.: Towards more realistic conceptual models of deep-water lobes: controls on net:gross and amalgamation ratio. In: Atlantic Ireland 2019, PIP-ISPSG, Dublin, pp. 41–42 (2019)

# Spatial Uncertainty in Pore Pressure Models at the Brazilian Continental Margin

Felipe Tajá C. Pinto, Krishna Milani, Leandro Guedes,
Luiz Eduardo S. Varella, Marcos Fetter, Marcus Santini, Thiago Lopes,
Vitor Gorne, Viviane Farroco, Attila L. Rodrigues, João Felipe C. L. Costa,
and Marcel A. A. Bassani

**Abstract** Accurate pore pressure models in wells are essential for ensuring the lowest cost and operational safety during exploration/development projects. This modeling requires the integration of several sources of information such as well data, formation pressure tests, geophysical logs, mud weight, geological models, seismic data, geothermal and sedimentation rate modeling. An empirical relationship between overpressure and compressional wave velocity is commonly applied to model the pore pressure. This deterministic approach does not allow uncertainty quantification and ignores other variables related to pore pressure. This paper presents a case study with real data to evaluate and quantify spatial pore pressure uncertainty. The exhaustive secondary variable came from the combination of seismic velocity and geothermal models. The methodology uses Sequential Gaussian Cosimulation with Intrinsic Collocated Cokriging. The results demonstrate the usefulness and applicability of the workflow proposed.

**Keywords** Pore pressure · SGSim · Intrinsic Collocated Cokriging

## 1 Introduction

The evaluation of pore pressure in an area of exploratory interest is important for optimizing the well design, such as correctly programming the casings, shoe setting, and appropriate well completion, to not waste resources and preserve safety all involved.

In these predictions, the oil industry applies the Eaton's Method [1], an empirical and deterministic method that infers the formation overpressure from compressional wave velocity, resistivity logs or even drilling parameters.

F. T. C. Pinto (✉) · K. Milani · L. Guedes · L. E. S. Varella · M. Fetter · M. Santini · T. Lopes · V. Gorne · V. Farroco
Petrobras SA, Rio de Janeiro, RJ, Brazil
e-mail: taja@petrobras.com.br

A. L. Rodrigues · J. F. C. L. Costa · M. A. A. Bassani
LPM/UFRGS, Porto Alegre, RS, Brazil

© The Author(s) 2023
S. A. Avalos Sotomayor et al. (eds.), *Geostatistics Toronto 2021*, Springer Proceedings in Earth and Environmental Sciences, https://doi.org/10.1007/978-3-031-19845-8_9

119

This modeling is interpretative since the modeler should propose a normal compaction trend, needed to detect overpressure zones. In many cases, the dataset needed to confirm the models is unavailable. Besides that, there is no way to quantify the variance or uncertainty of a model easily.

In order to reduce part of this interpretative aspect, and to be able to use more available data and give a statistical weight to the models, it was proposed to estimate the pore pressure in a portion of the eastern margin of the Brazilian coast via Sequential Gaussian Cosimulation [2, 3] with Intrinsic Colocated Cokriging [4].

## 2 Theoretical Foundations and Definitions

The used geostatistical modeling basically involves three steps that must be carefully evaluated to correctly calculate its probabilities. They are the variogram, the supersecondary variable [5] and the geostatistical simulation.

## 3 Data Presentation and Interpretation

As a region for the modeling, a recent exploratory frontier area was chosen due to its high-pressure domains and plane-parallel seismic facies.

Eight hundred eighty-three static pressure data from 24 wells were obtained, generating post-drilling models built from measured pressures, well logs and drilling reports.

To reduce the computational effort, we have delimited the modeling interval between the top of the Upper Campanian and the top of the Aptian. The histogram can be seen in Fig. 1.

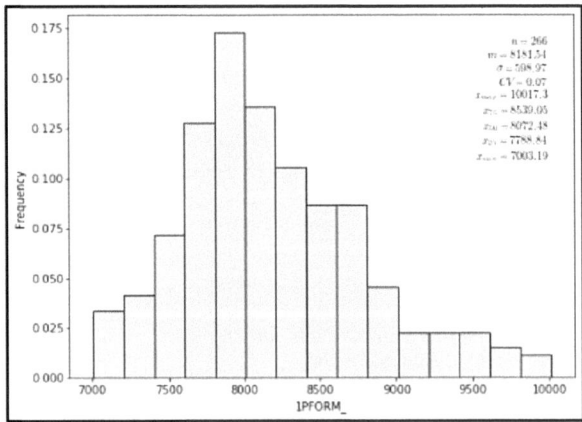

**Fig. 1** Pressure data (in psi) histogram from selected wells

The experimental variograms, as shown in Fig. 2, were obtained in the 0° and 90° azimuth directions and in the vertical direction respectively and were modeled from the Table 1. These variograms seem to be non-stationary, so the maximum search radii were limited.

Some 3D data from petroleum systems modeling were tested, such as sedimentation rate, organic matter transformation rate and porosity. Although some of these three-dimensional models correlate well with the primary data, their low resolution negatively impacted our final model. Therefore, these data were not included.

The supersecondary variable (Fig. 3) was obtained from the composition of the most recent models of seismic velocity (v) and temperature (T) according to the equation

$$S_s = -0.2v + 0.9T \tag{1}$$

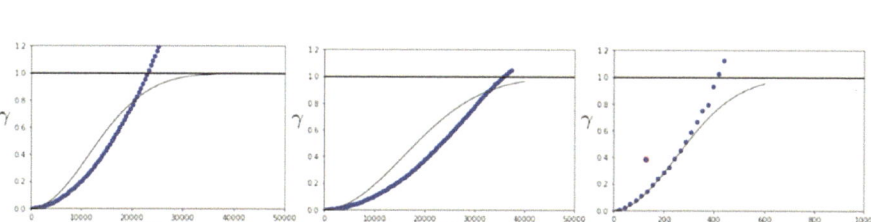

**Fig. 2** Experimental and modeled variograms. On the left, 0° azimuth with 35,000 m range. Centred is 90° azimuth with 50,000 m range and on the right, vertical with 900 m range

**Table 1** Modelled variogram parameters

| Model | Exponential |
|---|---|
| Nugget effect | 1% |
| Azimuth 0° (range) | 35,000 m |
| Azimuth 90° (range) | 50,000 m |
| Vertical (range) | 900 m |

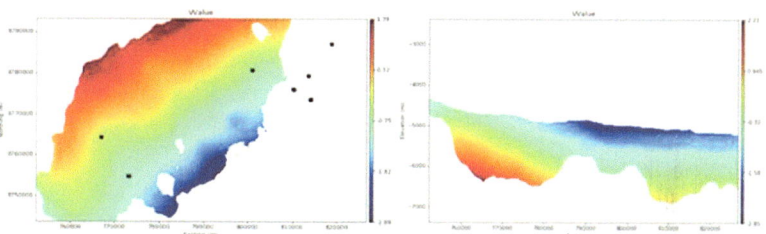

**Fig. 3** Sectional views of the supersecondary variable in-depth, on the left, and the x-axis, on the right

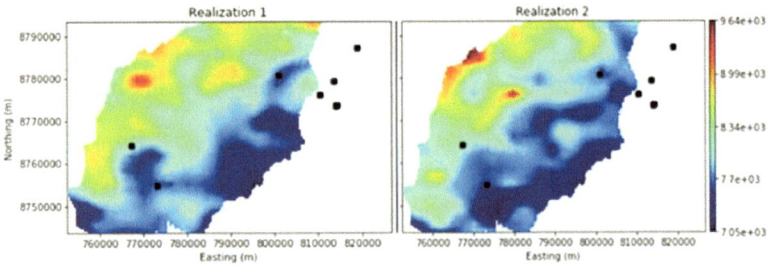

**Fig. 4** Depth slices of realizations 1 and 2, showing higher pressures in the NW region

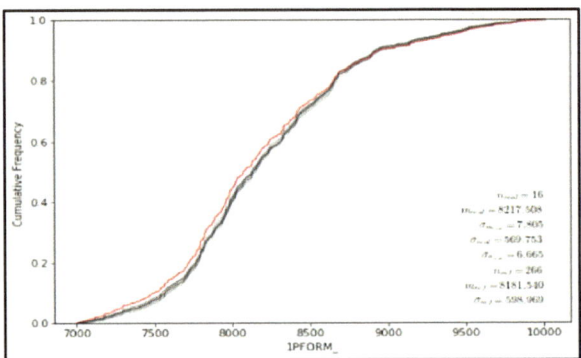

**Fig. 5** CDF reproduction over all realizations

Sixteen realizations were calculated in the Sequential Gaussian Co-Simulation with Intrinsic Collocated Cokriging. Some of these realizations can be seen in Fig. 4.

Figure 5 is possible to compare histograms of each realization, showing good agreement with the well data.

It was possible to obtain a basic statistic from all realizations, like mean values and standard deviation (Fig. 6) and percentiles like 10, 50 and 90 (Fig. 7) of the pore pressure in each cell grid.

Finally, the workflow presented permits extraction of the pressure log in any desired coordinate for a detailed evaluation Fig. 8.

## 4    Conclusions

The motivation of this work is to use cosimulation with intrinsic collocated cokriging to model a pore pressure–volume in a region of the Brazilian coast.

Post mortem pore pressure values in 24 recently drilled wells, limited between the Upper Campanian and Aptian horizons, were used as primary data. The secondary data were obtained from a composition of two variables: temperature modeled via

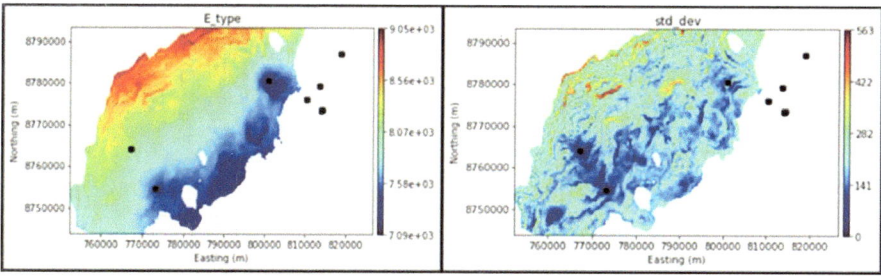

**Fig. 6** Mean value (left) and standard deviation (right) of the modeled pore pressure over all realizations

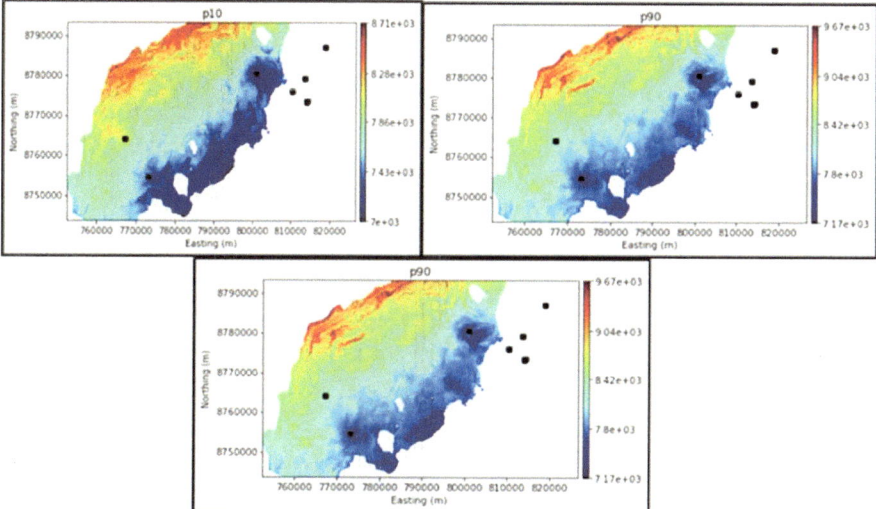

**Fig. 7** Percentiles 10, 50 and 90 of the modeled pore pressure

calibrated finite elements and the seismic velocity updated until the last well drilled in the area.

After the cosimulation, 16 realizations were obtained, and it was possible to see the wide spatial variability of pore pressure, with values within the range usually found in the region. These achievements honored the histogram and the original data and the modeled variogram, demonstrating the usefulness and applicability of the proposed workflow.

The presented geostatistical workflow demonstrates its applicability in future studies of pore pressure modeling, as they can generate different scenarios that are likely to be obtained at each realization, using a wider range of data available, and different well design strategies can be probabilistically defined.

**Fig. 8** Modeled pressure log
after simulation in a pair of
coordinates. All realizations,
percentiles 10 (blue), 50
(orange) and 90 (green) and
their standard deviation are
displayed

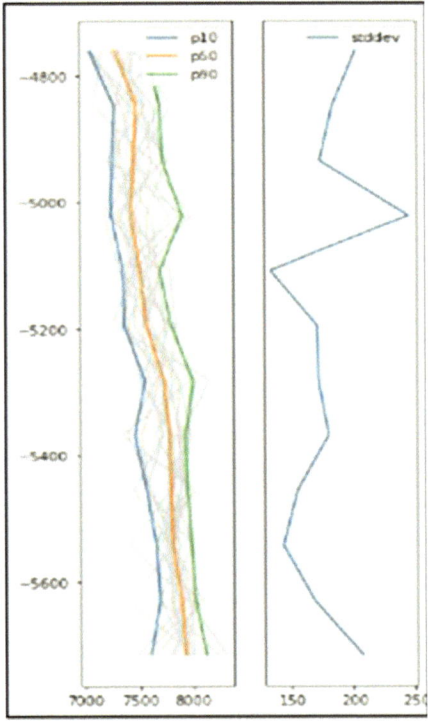

## 5  Benefits Promoted by This Work

One of the main issues in using the Eaton's method is determining a normal
compaction trend. This trend is quite interpretive, varying according to the modeler.
In order to reduce this subjectiveness, and mainly to include, in a statistical way,
other variables that may be related to pore pressure, the workflow in this article was
proposed using different exhaustive variables, from different models.

**Acknowledgements** The authors would like to thank Petrobras and LPM for the possibility of
developing this work.

## References

1. Eaton, B.A.: The Equation for Geopressure Prediction from Well Logs. SPE-5544-MS, 11p
   (1975)
2. Isaaks, E.H.: The application of Monte Carlo methods to the analysis of spatially correlated
   data. Ph.D. Thesis. Stanford University (1990)

3. Verly, G.W.: Sequential Gaussian Cosimulation: A Simulation Method Integrating Several Types of Information. In: Soares, A. (ed.) Geostatistics Tróia '92. Quantitative Geology and Geostatistics, vol. 5, pp. 543–554. Springer, Dordrecht (1993)
4. Babak, O., Deutsch, C.V.: An intrinsic model of Coregionalization that solves variance inflation in collocated Cokriging. Comput. Geosci. **35**(3), 603–614 (2009)
5. Babak, O., Deutsch, C.V.: Collocated Cokriging based on merged secondary attributes. Math. Geosci. **41**(8), 921–926 (2009)

# The Suitability of Different Training Images for Producing Low Connectivity, High Net:Gross Pixel-Based MPS Models

Deirdre A. Walsh⊙, Javier López-Cabrera, and Tom Manzocchi

**Abstract** Pixel-based multiple-point statistical (MPS) modelling is an appealing geostatistical modelling technique as it easily honours well data and allows use of geologically-derived training images to reproduce the desired heterogeneity. A variety of different training image types are often proposed for use in MPS modelling, including object-based, surface-based and process-based models. The purpose of the training image is to provide a description of the geological heterogeneities including sand geometries, stacking patterns, facies distributions, depositional architecture and connectivity. It is, however, well known that pixel-based MPS modelling has difficulty reproducing facies connectivity, and this study investigates the performance of a widely-available industrial SNESIM algorithm at reproducing the connectivity in a geometrically-representative, idealized deep-water reservoir sequence, using different gridding strategies and training images. The findings indicate that irrespective of the sand connectivity represented in the training image, the MPS models have a percolation threshold that is the same as the well-established 27% percolation threshold of random object-based models. A more successful approach for generating poorly connected pixel-based MPS models at high net:gross ratios has been identified. In this workflow, a geometrical transformation is applied to the training image prior to modelling, and the inverse transformation is applied to the resultant MPS model. The transformation is controlled by a compression factor which defines how non-random the geological system is, in terms of its connectivity.

**Keywords** Multiple-point statistics · Compression algorithm · Training image · Connectivity · Reservoir modelling

D. A. Walsh (✉) · J. López-Cabrera · T. Manzocchi
iCRAG and Fault Analysis Group, School of Earth Sciences, University College Dublin, Dublin, Ireland
e-mail: deirdre.walsh@icrag-centre.org

© The Author(s) 2023
S. A. Avalos Sotomayor et al. (eds.), *Geostatistics Toronto 2021*, Springer Proceedings in Earth and Environmental Sciences, https://doi.org/10.1007/978-3-031-19845-8_10

# 1  Introduction

Many deep-water lobe depositional sequences are characterized by laterally extensive, but finite, sand beds interbedded with continuous low permeability shales. Non-
random geological processes such as compensational stacking result in these systems
often having poor sand connectivity at high net:gross ratios (NTG). It is challenging
to reproduce this low connectivity in reservoir models, but it is important to do so if
the models are to be used to predict reservoir performance.

   The advantage of using a pixel-based method to model these sequences is that it
reproduces geological patterns while honouring conditioning well data. The connectivity of a number of training image (TI) inputs including object- and surface-based
models (OBM and SBM respectively) are investigated in this study and compared
to the connectivity of output MPS models built using the SNESIM algorithm [1].
The findings from these models, along with the connectivity characteristics of OBM
models, aided in the development of workflow for generating poorly connected MPS
models using a simple object-based TI followed by a grid transformation using the
compression algorithm.

# 2  Pixel-Based MPS Modelling with Common Training Images

The dimension of the model is 45 km × 45 km and 20 m thick. The target system
to be reproduced contains sand beds which are about 4.5 km wide and 0.5 m at their
thickest, NTG equal to about 40% and amalgamation ratio (AR) equal to about 10%.
This AR implies that 10% of the total area of sand bed bases should be connected to an
underlying sand bed. AR cannot be measured in an MPS model, so the connectivity
measure used to compare models is the proportion of total sand that is connected to
a vertical well at the centre of the model, which is about 5% in this case (Fig. 1).

   An unconditioned OBM is easily generated and works by placing user defined
objects within the model volume until the target NTG is reached. A downside of OBM
modelling is that are no under-defined constraints on the sand connectivity, and it is
clear in Fig. 1a that the object-based TI does not have the target low connectivity
since about 95% of the sand beds are connected to the well, either directly or via their
connections with different beds. Therefore, although the MPS model has reproduced
the connectivity of the TI, it does not have the low connectivity of the target system.
The second TI is a SBM generated using a number of depositional rules. These rules
control the placement of sand elements and shale interelements in a stratigraphic
order, and the probability that an element is capped by an interelement controls the
degree of sand amalgamation. An important feature of the TI that impacts the MPS
modelling is that the TI contains an irregular grid structure since each sand or shale
element is contained within an individual layer which is one grid cell thick. Hence,
although the TI represents the target system (Fig. 1b), the MPS algorithm produces

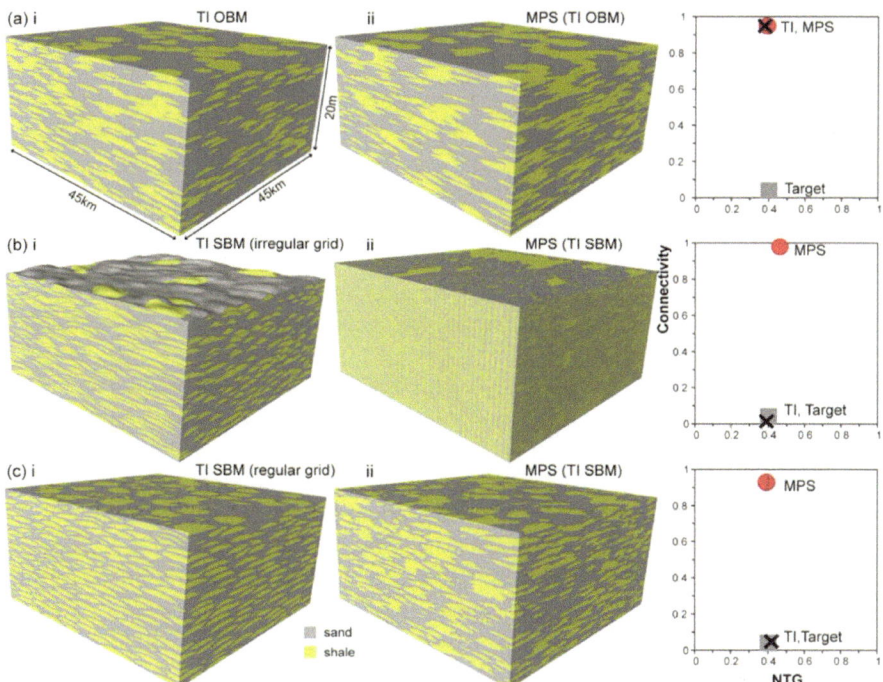

**Fig. 1** Summary of the various TIs (**a** OBM, **b** irregularly gridded SBM, **c** regularly gridded SBM) and the corresponding MPS models. The graphs show the connectivity in each case of the target system (■), the training image (x) and the MPS model (●)

very different model characteristics consisting of thin sand and shale layers resulting in complete sand connectivity. The final TI is a SBM constructed in a regular grid (Fig. 1c). This TI is a simple model generated with the same object-dimensions and 90% probability of shale drapes and therefore has low connectivity representative of the target model. However, again the MPS algorithm is unable to reproduce the low connectivity represented in the TI, and all of the sand is connected.

These results indicate that a simple workflow of constructing an accurate TI and applying this directly in MPS modelling is inappropriate for geological systems with non-random connectivity, such as the deep-water deposits considered. The only MPS model which reproduced the connectivity of the TI was the object-based case. The connectivity of MPS models generated by both OBM and SBM models are investigated in much greater detail [2] and it was found that these models follow the well-known percolation threshold seen in OBM whereby models with sand proportion greater than about 27% are well connected.

# 3   Pixel-Based Modelling with Low Connectivity

There are many modelling methods which allow the user to have control over the output connectivity including surface- or rule-based modelling, process-based and object-based methods. One such method, compression-based modelling, was developed originally in conjunction with OBM [3] and is based on the observation that the NTG and AR are equal in these models if all sand elements are of equal thickness. The compression method is a two-step process where modelling is undertaken with an initial low NTG (equal to the target AR) and cells containing sand and shale are expanded and compressed using the compression algorithm to reach the target NTG but preserved the low connectivity.

Combining the compression method with the MPS workflow requires that a low NTG OBM model is used as the TI input and, for the case considered here, it must equal the target low AR of 10% (Fig. 2a). The resulting MPS model reproduces the low connectivity of the TI (Fig. 2b). Then the compression algorithm is applied to the MPS model to rescale its NTG to the target value of 40% (Fig. 2c).

The properties of the training image, such as the thickness of the objects, are defined by the compression algorithm. This means that the decompression algorithm [4] is applied to the conditioning data prior to modelling. This is the inverse of the compression algorithm and ensures that the conditioning well data will be honoured when the compression algorithm is applied to the final model (Fig. 2d).

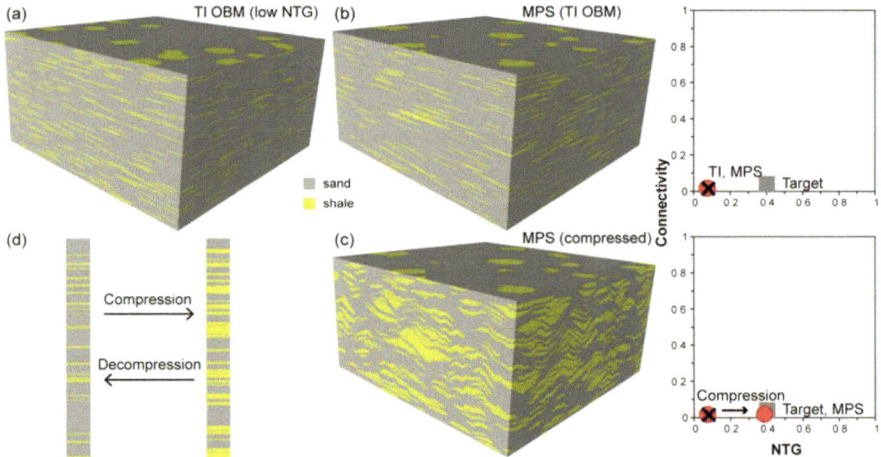

**Fig. 2** The compression-based MPS workflow

# 4 Summary

Several commonly used TIs have been investigated to examine both the suitability the TI to honour low connectivity at high NTG, and the ability of the MPS algorithm to reproduce the connectivity represented in the TI (Table 1). Although the object-based TI does not contain low connectivity at NTG values greater than 27%, MPS models based on these TIs reproduce accurately their connectivity. In contrast, the two surface-based TIs have the target low connectivity, but this connectivity is not reproduced in the MPS models. A detailed investigation indicates that MPS models follow the same relationship between NTG and connectivity as object-based models when built using surface- or object-based TIs [2]. This equivalence provides the basis for a new workflow in which MPS models are built using OBM TIs, and then rescaled to the target NTG using the compression algorithm. The conditioning well data is incorporated by using the inverse transformation prior to the MPS modelling step. The final MPS model honours the well data and contains independently user defined NTG and connectivity.

**Table 1** Summary of the connectivity of a number of TIs (OBM and SBM) and the resulting MPS models generated with or without the compression algorithm (CA)

| Type of TI | Modelling workflow | TI has target NTG | TI has target connectivity? | Final model honours target connectivity? | Comments |
|---|---|---|---|---|---|
| OBM | MPS | Yes | No | No | The MPS model reproduced the TI, but this does not have the target low connectivity |
| SBM[a] | MPS | Yes | Yes | No | The MPS algorithm cannot resolve the irregular grid structure of the TI and therefore is unable to reproduce its connectivity |
| SBM | MPS | Yes | Yes | No | The MPS algorithm does not reproduce the low connectivity of the TI |
| OBM | MPS then CA | No | Yes | Yes | The target low connectivity at high NTG is achieved via a geometrical transformation of a low NTG MPS model using the CA |

[a] This TI has an irregular grid structure

**Acknowledgements** This presentation has emanated from research supported in part by a research grant from Science Foundation Ireland (SFI) under Grant Number 13/RC/2092_P2 and is co-funded by PIPCO RSG and its member companies. Schlumberger are thanked for the provision of an academic Petrel license.

# References

1. Strebelle, S.: Conditional simulation of complex geological structures using multiple-point statistics. Math. Geol. **34**, 1–21 (2002)
2. Walsh, D.A., Manzocchi, T.: Connectivity in pixel-based facies models. Math. Geosci. **53**, 415–435 (2021)
3. Manzocchi, T., Zhang, L., Haughton, P.D.W., Pontén. A.: Hierarchical parameterization and compression-based object modelling of high net:gross but poorly amalgamated deep-water lobe deposits. Pet. Geosci. **26**, 545–567 (2020)
4. Walsh, D.A., Manzocchi, T.: A method for generating geomodels conditioned to well data with high net:gross ratios but low connectivity. Marine Pet. Geol. **129** (2021)

# Probabilistic Integration of Geomechanical and Geostatistical Inferences for Mapping Natural Fracture Networks

Akshat Chandna and Sanjay Srinivasan

## 1 Introduction

Estimation of a reservoir's production potential, well placement and field development depends largely on accurate modeling of the existing fracture networks. However, there is always significant uncertainty associated with the prediction of spatial location and connectivity of fracture networks due to lack of sufficient data to model them. Therefore, stochastic characterization of these fractured reservoirs becomes necessary.

Two-point statistics-based algorithms are inadequate for describing complex spatial patterns such as branching and termination of fractures described by the joint variability at multiple locations at a time [7]. Constraining the models to multiple point statistics (MPS) is necessary for producing maps that are able to accurately predict termination and intersection of the fractures without having to separate the fracture sets on the basis of their chronological evolution that may be difficult due to sparse data [5, 7]. In general, MPS is fast and robust, and superior to the traditional two point statistics while realistically reproducing the complex curvilinear geologic structures as well as integrating different data sets [6]. Conventional MPS algorithms depend on a well-defined spatial template to capture multi-scale features. Gridded domains are inefficient and tend to interrupt the spatial connectivity of the fractures. The ideas of non-gridded TIs, templates and simulated images put forth by Erzeybek (2012) are extended in a fast, robust and easily scalable MPS algorithm that utilizes self-adjusting and automatic template selection based on the configuration of conditioning data around the simulated node [1]. At the initial stages of modeling, the template identifies the coarse scale pattern and as the modeling progresses, patterns over fine and finer scales are reproduced.

A. Chandna (✉) · S. Srinivasan
Pennsylvania State University, State College, PA 16801, USA
e-mail: akshatch@gmail.com

© The Author(s) 2023
S. A. Avalos Sotomayor et al. (eds.), *Geostatistics Toronto 2021*, Springer Proceedings in Earth and Environmental Sciences, https://doi.org/10.1007/978-3-031-19845-8_11

Geomechanical modeling of fractures is another widely popular approach to map fracture networks constrained to the physics of the reservoir such as far field stress conditions, presence of faults and other geological structures and local stress effects of nearby fractures. However, development of full reservoir scale physics model involving material heterogeneities beyond a length scale of 1 km involves extensive computation costs and time. It is also imperative that the uncertainties in reservoir parameters are accounted for in the prediction of reservoir performance [5]. Inferring geomechanical rules for fracture propagation in a probabilistic sense is necessary to represent the uncertainty. This is achieved using Machine Learning (ML) approaches trained on high-fidelity small scale FDEM models that predict fracture propagation pathways given a set of physics-based parameters [2].

A statistics based approach does not consider the physical processes guiding fracture propagation and a geomechanics based approach may not honor the fracture statistics observed from other auxiliary sources such as outcrop images. Therefore, amalgamation of the MPS and geomechanics based approaches is ideal for producing fracture networks, constrained to both reservoir physics and reservoir statistics. This research presents a paradigm for integration of information obtained from a stochastic simulation algorithm and geomechanics based algorithm using the Tau model proposed by Journel [4] that utilizes the concept of permanence of ratios.

## 2 MPS Algorithm in Classification Framework

A new and improved stochastic simulation technique based on MPS presented by Chandna (2019) is shown to improve upon the shortcomings of the classical MPS algorithms [1]. It is able to generate the desired fracture patterns without relying on any grid, either for the template or simulated image. This algorithm employs self-manipulating templates to include the specified maximum number of nodes in the vicinity of the simulation node, thereby eliminating the need to predefine templates based on visual observation and initial analysis of the training image (TI) that generally fail to capture either the small or the large-scale features unless multi-grid simulations are performed. It also circumvents calculating multiple point histograms since the algorithm operates on the principle of direct sampling [3]. In direct sampling, the pattern identified using the data configuration around the simulation node (rather than using a fixed spatial template) is searched in the TI and corresponding to the first instance of a match, the outcome at the simulation location in the TI is directly extracted and applied to the simulation. This results in more computational efficiency as the entire TI need not be searched for the calculation of the number of occurrences of the desired pattern.

The ML based geomechanical simulation algorithm outputs probabilities of the propagation of a fracture tip in each of 8 angles classes formulated by dividing the circular region around the fracture tip in 8 equal sectors of angle $\pi/4$ centered at the fracture tip (Fig. 1a). But the MPS algorithm presented by Chandna [1] is regression based and outputs discrete angles of propagation for each simulated fracture tip. For

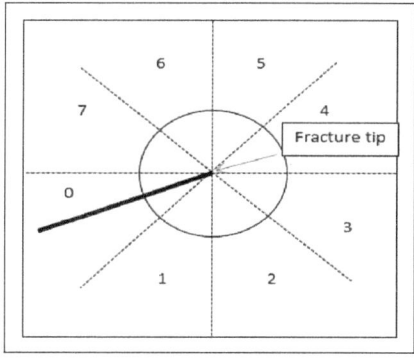
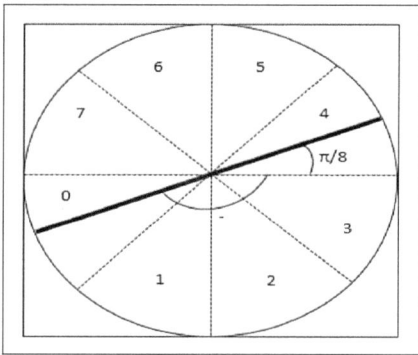

(a) A propagating fracture

(b) An existing fracture in a TI showing conjugate angle classes (for example class 0 and 4)

**Fig. 1** Angle classes 0–7, around a fracture denoted by a thick black line [2]

integrating the probability obtained from geomechanical modeling and that from MPS simulation, multiple angles of propagation are simulated for the same fracture tip, which can be binned together in these angle classes and their counts can be used to estimate the probability of simulation of each angle class. The first modification made to the algorithm constitutes outputting probabilities of angle classes of propagation for the node being simulated ($P(\theta_i)$ where $i$ can range from 0 to 7) instead of discrete angles. Since, the MPS algorithm is based on direct sampling of the propagation angle from the TI, one option is to sample the same TI multiple times using the same template pattern to obtain multiple propagation angle classes. Each propagation angle could be different due to random order of simulation of initial flaws. These propagation angles can then be used to calculate the probability of observing a particular angle class. This can be represented as:

$$P(\theta_i) = \frac{Number\ of\ times\ \theta_i\ is\ sampled\ from\ the\ TI}{Total\ number\ of\ TI\ samplings} \tag{1}$$

A better option would be to sample multiple TIs using the same template to account for the uncertainty in the TI itself i.e.:

$$P(\theta_i) = \frac{Number\ of\ TIs\ from\ which\ \theta_i\ is\ sampled}{Total\ number\ of\ TIs} \tag{2}$$

The second approach is adopted in this research.

## 3 Combination of Probabilities

The combination of contributions of data events B and C from different sources to predict the probability of a desired event A is one of the most common issues faced in data sciences and most importantly in earth sciences. This conditional probability of an unknown event A occurring given two data events B and C from different sources can be expressed as $P(A|B, C)$. Commonly, this conditional probability is estimated assuming some level of independence of the data events B and C. However, such an assumption leads to non-robust algorithms with questionable prediction accuracy. Journel (2002) proposed an alternate probability integration paradigm based on the permanence of ratios which assumes that the relative contribution of any one data event to the occurrence of an outcome is independent of the relative contributions of all other data events [4]. This integration algorithm is used in the current research as a means to combine probabilities of simulated angle classes generated from the two sources: the MPS algorithm and the geomechanical simulation algorithm. The underlying assumptions of the probability integration algorithm are that the prior probability of the outcome data event $(P(A))$ and the probability of the outcome data event given the source data event for each of the source data event, $(P(A|B)$ and $P(A|C))$ can be evaluated. The permanence of ratio hypothesis breaks down the information from each source but recombines these elemental probabilities using a tau $(\tau)$ parameter(s). The tau parameters actually explore the redundancy between different data.

For every step of propagation of a fracture tip, probabilities of propagation along all angle classes 0 to 7 are obtained from two different sources: MPS based algorithm (Sect. 3) and geomechanics based algorithm (Sect. 2). Let $P(\theta)$ be the probability of the data event: simulation of angle class $\theta_i$ where i ranges from 0 to 7. Let $P(\theta_i|B)$ be the probability of simulation of angle class $\theta_i$ obtained from the MPS simulation and $P(\theta_i|C)$ be the probability of angle class $\theta_i$ obtained from geomechanical simulation. The desired probability of occurrence of angle class $\theta_i$ given the probabilities of its joint occurrence from the MPS and geomechanical simulations, $P(\theta_i|B, C)$ can then be evaluated using the concept of permanence of ratios:

$$P(\theta|B, C) = \frac{1}{1 + b(\frac{c}{a})^\tau} \tag{3}$$

$$a = \frac{1 - P(\theta)}{P(\theta)} = \frac{P(\tilde{\theta})}{P(\theta)} \in [0, +\infty] \tag{4}$$

$$b = \frac{1 - P(\theta|B)}{P(\theta|B)} = \frac{P(\tilde{\theta}|B)}{P(\theta|B)} \in [0, +\infty] \tag{5}$$

$$c = \frac{1 - P(\theta|C)}{P(\theta|C)} = \frac{P(\tilde{\theta}|C)}{P(\theta|C)} \in [0, +\infty] \tag{6}$$

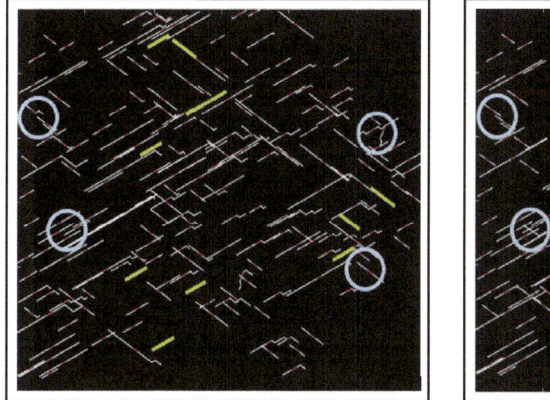

(a) Simulated image using geomechanics algorithm

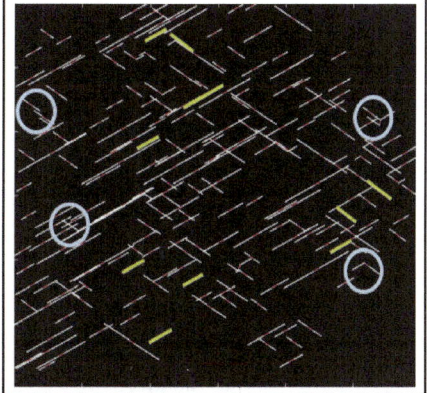

(b) Simulated image using MPS algorithm

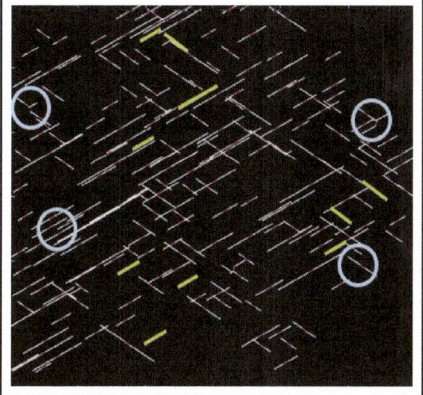

(c) Simulated image using integration algorithm

**Fig. 2** Comparison of final simulated images using the geomechanics, MPS and integration algorithms. Blue circles indicate highlighted areas of interest.

The contribution of data event C is manipulated and tuned using the $\tau$ parameter. For $\tau = 0$, the contribution of data event C is completely ignored and the contribution is increased or decreased depending on if $\tau > 1$ or $< 1$ respectively.

For demonstrating the model for integration of probabilities obtained from MPS in classification framework and reduced order machine learning based geomechanical simulation model, for propagation of every node, the integrated probability distribution over the angle classes is obtained and the node is propagated in the direction of the angle class with maximum probability of occurrence. The prior uncertainty $P(\theta)$ that results in the ratio $a$ over angle class $\theta_i$ is assumed to be uniform and is updated based on the incremental contribution from the MPS and geomechanical simulation algorithm.

Figure 2 shows a comparison of the final simulated fracture maps generated using the geomechanics, the MPS and the integration algorithm. $\tau$ is assumed to be 1 for this simulation, implying that the relative contributions of the MPS and geomechanics algorithm to the knowledge of final predicted propagation angle, are assumed to be independent of each other. Few areas are highlighted by blue circles that show the effect of the integration algorithm on the simulated fracture maps after combining information from the two individual sources: MPS and geomechanics. Most of these correspond to hooking like pattern of the fractures. In three of the highlighted areas, statistics derived from the TIs by the MPS algorithm facilitate hooking of the fractures. However, due to the stress regimes developed around these fractures, the geomechanics algorithm predicts propagation of these fractures without any hooking with a high probability. After combining the probabilities over all possible angle classes that can be simulated, the integration algorithm simulated an angle propagation class that did not favor hooking of the fractures. Similarly, in one of the highlighted areas, hooking is favored by the integration algorithm due to a stress regime caused by possibly significant fracture interactions. In general, due to the coarse angle classes used in geomechanical simulation, a number of kinks are observed in geomechanically simulated fractures. These kinks arise as fractures tend to merge or diverge from other fractures according to the progressively changing stress states around propagating fracture tips. However, if such phenomena is not observed in the TIs, the MPS algorithm does not simulate these features. It then depends on the incremental contribution of the MPS and the geomechanical information to the knowledge of the final predicted propagation angle class to determine if such features would be observed in the final simulated images. In this case, such features are not simulated by the integration algorithm due to lower contribution of the geomechanics algorithm towards generation of propagation angle classes that may describe the development of these features.

**Acknowledgements** The authors would like to acknowledge the support and funding from Penn State Initiative for Geostatistics and Geo-Modeling Applications (PSIGGMA) and the member companies.

# References

1. Chandna, A., Srinivasan, S.: Modeling natural fracture networks using improved geostatistical inferences. Energy Proc. **158** 6073–6078 (2019). Innovative Solutions for Energy Transitions
2. Chandna, A., Srinivasan, S.: Mapping natural fracture networks using geomechanical inferences from machine learning approaches. Comput. Geosci. **26**(06) (2022)
3. Gregoire, M., Philippe, R., Julien, S.: The direct sampling method to perform multiple-point geostatistical simulations. Water Resour. Res. **46**(11) (2010)

4. Journel, A.G.: Combining knowledge from diverse sources: an alternative to traditional data independence hypotheses. Math. Geol. **34**(5), 573–596 (2002)
5. Liu, X., Srinivasan, S.: Field scale stochastic modeling of fracture networks-combining pattern statistics with geomechanical criteria for fracture growth. In: Leuangthong, O., Deutsch, C.V. (eds.) Geostatistics Banff 2004, pp. 75–84. (2005)
6. Strebelle, S.: Conditional simulation of complex geological structures using multiple-point statistics. Math. Geol. **34**(1), 1–21 (2002)
7. Strebelle, S.B., Journel, A.G.: Reservoir modeling using multiple-point statistics. In: SPE Annual Technical Conference and Exhibition (2001)

# Mining

# Artifacts in Localised Multivariate Uniform Conditioning: A Case Study

Oscar Rondon and Hassan Talebi

**Abstract** Localised Multivariate Uniform Conditioning (LMUC) is a technique designed for spatially locating Selective Mining Unit (SMU) grades derived using Multivariate Uniform Conditioning (MUC) for the assessment of recoverable resources. LMUC has the advantage of producing SMU estimates conforming to the MUC panel-specific grade-tonnage curves while preserving the spatial grade distribution at the selective mining level. However, LMUC results have two severe artifacts. This paper documents both artifacts using four grades from a large nickel–cobalt laterite deposit in Western Australia.

**Keywords** Multivariate uniform conditioning · Recoverable resources · Multivariate grade localisation

## 1 Introduction

Uniform Conditioning (UC) uses a set of predetermined cut-offs to estimate the grade, metal and tonnage of SMUs inside a panel using the estimated panel grade [1, 2]. These estimates are often not practical for many mining studies because of the limited capacity for visualising the likely spatial grade distribution at the SMU scale. Post-processing using localised uniform conditioning (LUC) allows to localise the SMU grades in such a way that SMU blocks have grade-tonnage relationships matching the UC results [3]. This is useful, for instance, for open pit mining studies that require evaluating the economics of a project.

Deraisme et al. [4] extended the concept of LUC to the multivariate case. The key idea is to analogously use the panel-specific recovery curves derived from MUC

O. Rondon (✉)
Snowden Optiro, Level 19, 140 St Georges Terrace, Perth 6000, Australia
e-mail: oscar.rondon@snowdenoptiro.com

H. Talebi
CSIRO Deep Earth Imaging FSP, 26 Dick Perry Avenue, Kensington, WA 6151, Australia

CSIRO Mineral Resources, 26 Dick Perry Avenue, Kensington, WA 6151, Australia

School of Science, Edith Cowan University, Joondalup, WA 6027, Australia

© The Author(s) 2023
S. A. Avalos Sotomayor et al. (eds.), *Geostatistics Toronto 2021*, Springer Proceedings in Earth and Environmental Sciences, https://doi.org/10.1007/978-3-031-19845-8_12

[5] to localise the results. The process is straightforward and has practical merits. However, it produces two artifacts that suggest the multivariate localisation is flawed. The first one corresponds to unrealistic linear-like patterns in the scatterplots between localised attributes. This artifact can be partially detected in Fig. 7 of [4] and it is clearly seen in Fig. 8 of [6]. The second artifact relates to the discrepancy between the expected theoretical correlation at the SMU support derived from MUC and the corresponding one between localised attributes. In this article, both artifacts are discussed using data from a large nickel–cobalt laterite deposit in Western Australia as a case study.

The article is organised as follows: firstly, a summary of MUC and LMUC are presented where the concepts and notations used are introduced. Secondly, a summary of the nickel–cobalt laterite deposit characteristics along with details of the LMUC application are presented. Thirdly, the case study results are discussed. Lastly, conclusions and recommendations are drawn on the bases of the findings.

## 2   Multivariate Uniform Conditioning and LMUC

MUC represents a most challenging problem in Geostatistics. For a set of multivariate attributes $(Z_1, Z_2, \ldots, Z_N)$ the problem consists in estimating recoverable resources inside a mining panel $V$ using the set of multivariate estimated panel grades $(Z_1^*(V), Z_2^*(V), \ldots, Z_N^*(V))$ where each estimate is assumed to be conditionally unbiased.

The MUC model proposed by Deraisme et al. [5] proceeds first by independently applying the discrete Gaussian method [2] to each attribute $Z_i$, $i = 1, \ldots, N$ which allows expressing the grades $Z_i(x)$ at point support, $Z_i(v)$ at SMU support, and the estimated grades $Z_i^*(V)$ at panel support as functions of standard Gaussian variables $Y_i(x)$, $Y_{iv}$ and $Y_{iV}^*$ respectively given by

$$Z_i(x) = \Phi_i(Y_i(x)) = \sum_{n \geq 0} \phi_{in} H_n(Y_i(x)), \qquad (1)$$

$$Z_i(v) = \Phi_{iv}(Y_{iv}) = \sum_{n \geq 0} \phi_{in} r_i^n H_n(Y_{iv}), \qquad (2)$$

and

$$Z_i^*(V) = \Phi_{iV}^*(Y_{iV}^*) = \sum_{n \geq 0} \phi_{in} s_i^n H_n(Y_{iV}^*), \qquad (3)$$

where $H_n$ are the Hermite polynomials and $\phi_{in}$ are the corresponding coefficients with $r_i$ and $s_i$ the variance correction factors for SMU and panel support respectively. Here, the same number of Hermite polynomials is being used for all attributes. This is just a minor restriction required for computation of theoretical values during MUC such as the covariance between the SMU support grades in Eq. 6. The MUC model is entirely specified once the variance correction factors $r_i$ and $s_i$ and the correlations between the equivalent Gaussian variables are calculated. The former is obtained by inverting Eqs. 4 and 5

$$Var(Z_i(v)) = \sum_{n \geq 1} \phi_{in}^2 r_i^{2n} \tag{4}$$

$$Var(Z_i^*(V)) = \sum_{n \geq 1} \phi_{in}^2 s_i^{2n} \tag{5}$$

while the latter at the SMU support $v$ by inverting Eq. 6

$$Covar(Z_i(v), Z_j(v)) = \sum_{n \geq 1} \phi_{in} \phi_{jn} r_i^n r_j^n \rho_{ij}^n, \tag{6}$$

where $\rho_{ij}$ corresponds to the correlation between $Y_{iv}$ and $Y_{jv}$. This implies that the $Corr(Z_i(v), Z_j(v))$ between any two different attributes $Z_i$ and $Z_j$ at the SMU $v$ support can be explicitly computed using Eqs. 4 and 6. Through a set of conditional independence assumptions and the use of a master or anchor attribute for identifying whether or not a given SMU is above a cut-off, the MUC model effectively reduces the multivariate case to a bivariate one and the values of the remaining parameters required to completely specify the model can be calculated. Selection of the master attribute is of prime importance because the MUC model is driven by the correlations between the master and all other attributes while other correlations can, at most, be partly inferred through their corresponding relations with the main attribute. The reader is referred to Deraisme et al. [5] for further details.

Deraisme et al. [4] extended the concept of LUC to the multivariate case. Using the panel-specific tonnage curve of the master attribute along with the panel-specific metal curves of all attributes derived from MUC, the corresponding grades can be localised as done in LUC. The ranking is though entirely based on the direct estimate of the master attribute at the SMU support.

## 3   Case Study Presentation and Results

The data used for the case study comes from the Murrin Murrin East (MME) nickel–cobalt laterite deposit in Western Australia. The main mineralised body is approximately 1,500 m long, 600 m wide, and 30 m thick (Fig. 1). Surficial chemical weathering of ultramafic rocks resulted in the formation of nickel–cobalt laterites. Mineralisation occur as laterally extensive and undulating blankets with strong vertical zonation [7]. The attributes of interest include Co(%), Fe(%), Mg(%), and Ni(%) which was used as the master attribute. In total, 20,690 samples of 1 m length from 926 RC holes comprise the database for this study. The RC holes are approximately

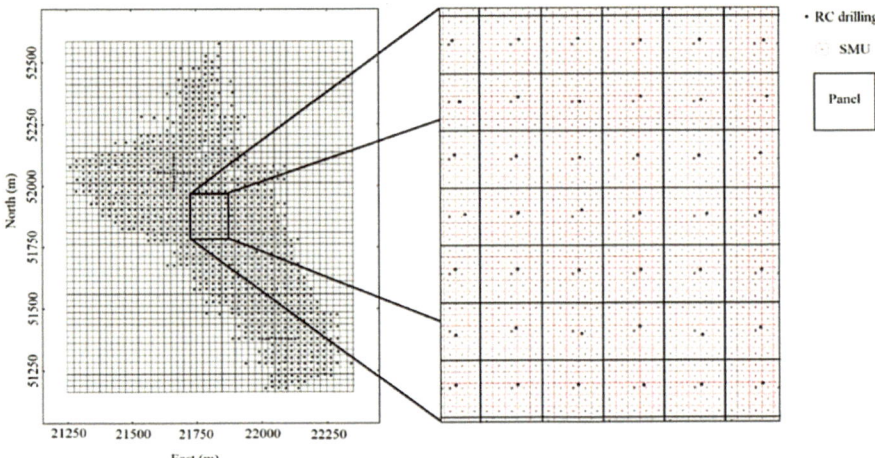

**Fig. 1** Plan view of the MME deposit, RC drillings (black dots), SMUs (red squares), and panels (black squares)

**Table 1** Classical descriptive statistics (left) and correlation matrix (right) of the selected attributes

|    | Min   | Max    | Mean   | Var     | Fe     | Mg     | Ni    | Co     |
|----|-------|--------|--------|---------|--------|--------|-------|--------|
| Fe | 0.400 | 61.500 | 20.463 | 159.510 | 1.000  | −0.755 | 0.008 | 0.198  |
| Mg | 0.015 | 26.800 | 6.818  | 43.230  | −0.755 | 1.000  | 0.119 | −0.168 |
| Ni | 0.004 | 4.070  | 0.715  | 0.250   | 0.008  | 0.119  | 1.000 | 0.583  |
| Co | 0.001 | 2.270  | 0.054  | 0.010   | 0.198  | −0.168 | 0.583 | 1.000  |

located on a square grid of 25 m × 25 m. The panel and SMU dimensions are 25 m × 25 m × 2 m and 5 m × 5 m × 1 m respectively. Consequently, each panel contains 50 SMUs (Fig. 1). Descriptive statistics and correlations between attributes are shown in Table 1.

## 3.1 Global and Local Scatterplots

Figure 2 shows the global scatterplot between localised Ni and Co grades along with the same scatterplot for SMUs inside five different panels selected randomly across the deposit. Globally, the localised grades at the SMU support seem to reproduce the input correlation between Ni and Co. However, locally, i.e., for SMUs in a panel, the results are drastically different. The correlation of 0.583 between Ni and Co becomes almost perfect for SMUs inside the five panels with clear linear-like patterns. This artifact is a direct consequence of using a master attribute to localise the SMU grades as it will be shown next.

**Fig. 2** Scatterplot between
input Ni and Co grades (top)
and corresponding localised
grades (bottom). Coloured
lines are the scatterplots
between localised grades for
SMUs inside a panel for five
panels selected randomly
across the deposit

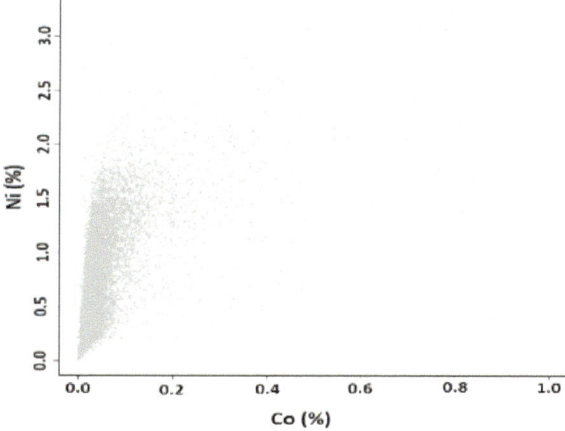

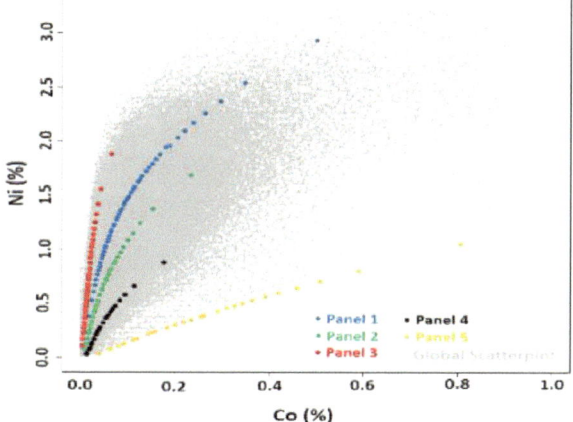

## 3.2 Correlation Between Localised Attributes

The multivariate localisation imposes relationships between attributes that do not
necessarily guarantee the reproduction of the global expected theoretical correlations
$Corr\big(Z_i(v), Z_j(v)\big)$ given by the MUC model.

Without loss of generality, let $Z_1$ be the master attribute and $(Z_1, Z_2, \ldots, Z_N)$
the localised attributes inside a panel $V$. The average grade $M_i$ for $Z_i$ is the ratio of
the amount of metal $Q_i$ and tonnage $T_1$ for $i = 1, 2, \ldots, N$. Therefore, the average
grades $M_i i \geq 2$ are related to $M_1$ by Eq. (7)

$$M_i = \frac{Q_i}{Q_1} M_1 \qquad (7)$$

**Table 2** Comparison of expected theoretical (low diagonal) and localised attributes (upper diagonal) correlations at the SMU support

|      | Co     | Fe     | Mg     | Ni     |
|------|--------|--------|--------|--------|
| Co   |        | 0.228  | −0.038 | 0.735  |
| Fe   | 0.214  |        | −0.825 | −0.046 |
| Mg   | −0.167 | −0.772 |        | 0.321  |
| Ni   | 0.583  | −0.003 | 0.148  |        |

Equation (7) implies that when the master attribute $Z_1$ falls within a grade class $[a, b)$ during the localisation, the secondary attribute $Z_i$ is calculated as

$$Z_i = \frac{Q_i(a) - Q_i(b)}{Q_1(a) - Q_1(b)} Z_1 \quad i \geq 2 \tag{8}$$

where $Q_i(a)$ and $Q_i(b)$ are the panel-specific metal curves for attribute $Z_i$ evaluated at $a$ and $b$ respectively. This shows why after the localisation, the scatterplots between the master and all other localised attributes exhibit linear-like patterns. Equation (8) also shows that localised secondary attributes are implicitly related by

$$Z_i = \frac{Q_i(a) - Q_i(b)}{Q_j(a) - Q_j(b)} Z_j \quad i \neq j \tag{9}$$

A comparison of the expected theoretical and localised attributes correlations is provided in Table 2.

Moreover, localised secondary attributes may have spurious results. For instance, although Fe and Mg have negative correlation, the scatterplot between the corresponding localised grades inside a panel exhibit positive correlations (Fig. 3).

# 4 Conclusions

The current implementation of MUC reduces a multivariate problem to a bivariate one through the nomination of a master or anchor attribute for identifying SMUs that are above a cut-off of interest. Although this seems reasonable for MUC applications, localising the MUC results produces two severe artifacts. These are a direct consequence of using the master attribute along with the MUC results to guide the localisation.

LMUC results exhibit linear-like patterns not consistent with the characteristics of the input data. Those correspond to the scatterplots of localised attributes within a panel. The consequence of this is that the correlation between localised attributes is drastically different at the global, i.e., across the mineralised domain under study, and local scale, i.e., inside a mining panel.

**Fig. 3** Scatterplot between input Fe and Mg grades (top) and corresponding localised grades (bottom). Coloured lines are the scatterplot between corresponding localised grades for SMUs inside the same panels used in Fig. 2

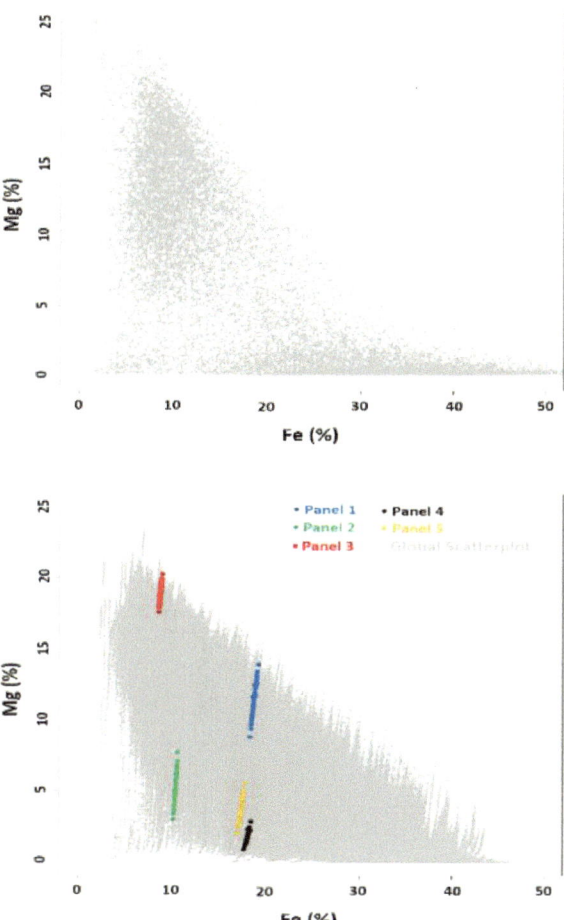

**Acknowledgements** The authors would like to thank Professor Ute Mueller for her help improving the manuscript. The second author acknowledges the funding and support from CSIRO's Deep Earth Imaging Future Science Platform.

# References

1. Chilès, J.-P., Delfiner, P.: Geostatistics: Modeling Spatial Uncertainty, 2nd edn. Wiley, New York (2012)
2. Rivoirard, J.: Introduction to Disjunctive Kriging and Non-linear Geostatistics. Oxford Press, Claredon (1994)
3. Abzalov, M.Z.: Localised uniform conditioning (LUC): A new approach for direct modelling of small blocks. Math. Geol. **38**, 393–411 (2006)

4. Deraisme, J., Assibey-Bonsu, W., Baafi, E.Y., et al.: Localised uniform conditioning in the multivariate case-an application to a porphyry copper gold deposit. In: Proceedings of the 35th APCOM Conference (2011)
5. Deraisme, J., Rivoirard, J., Carrasco, P.: Multivariate uniform conditioning and block simulations with discrete gaussian model: application to chuquicamata deposit. In: VIII International Geostatistics Congress, GEOSTATS, pp. 69–78 (2008)
6. Deraisme, J., Assibey, W.: Comparative study of localized block simulations and localized uniform conditioning in the multivariate case. In: Abrahamsen, P., Hauge, R., Kolbjørnsen, O. (eds.) Geostatistics Oslo 2012, pp. 309–320. Springer, Netherlands Dordrecht (2012)
7. Murphy, M.: Geostatistical Optimisation of Sampling and Estimation in a Nickel Laterite Deposit. Edith Cowan University (unpublished) (2003)

# Methodology for Defining the Optimal Drilling Grid in a Laterite Nickel Deposit Based on a Conditional Simulation

Claudia Mara Sperandio Neves, João Felipe Coimbra Leite Costa, Leonardo Souza, Fernando Guimaraes, and Geraldo Dias

**Abstract** In mining projects, the confidence in an estimate is associated with the quantity and quality of the available information. Thus, the closer the data to the targeted location, the smaller the error associated with the estimated value. In the advanced stages of a project (i.e. the pre-feasibility and feasibility phases), it is usual to take samples derived from drillings. Since sampling and chemical analysis involve high costs, it is essential that these costs contribute to a reduction in the uncertainty of estimation. This paper presents a workflow for a case study of a lateritic nickel deposit and proposes a methodology to address the issue of optimising the drilling grid based on uncertainty derived from Gaussian conditional geostatistical simulations. The usefulness of the proposed workflow is demonstrated in terms of saving time and money when selecting a drill hole grid.

**Keywords** Conditional simulation · Optimal drill spacing · Estimation uncertainty

## 1 Introduction

Although infill drilling is mandatory to reduce the uncertainty in estimated figures, the drilling grid is frequently selected without reference to any geostatistical criteria to support or optimise the locations of these additional drill holes. It is also important to determine the point at which these additional drillings become irrelevant in terms of reducing the associated error, since in this case, infill drilling will only result in temporal and financial losses.

C. M. S. Neves (✉)
Vale, Rio de Janeiro, Brazil
e-mail: sebastian.avalos@queensu.ca; claudianeves82@gmail.com

J. F. C. L. Costa
Mining Engineering Department, Federal University of Rio Grande, Rio Grande Do Sul, Brazil

L. Souza
Talisker Exploration Services, Toronto, Canada

F. Guimaraes · G. Dias
Anglo American, Belo Horizonte, Minas Gerais, Brazil

© The Author(s) 2023
S. A. Avalos Sotomayor et al. (eds.), *Geostatistics Toronto 2021*, Springer Proceedings in Earth and Environmental Sciences, https://doi.org/10.1007/978-3-031-19845-8_13

The use of geostatistics can assist in the construction of models that have more than one variable, which can be estimated using ordinary kriging [1] or any other technique derived from it. However, this procedure does not make it possible to access the correct uncertainty associated with the estimated value, since the variance in the interpolated values is less than the variance of the original data. The limitations related to the use of kriging variance as a measure of uncertainty have been extensively discussed in the literature [2, 3]. The kriging variance considers only the spatial distribution of the samples, and does not take into account their values and local variability [4, p. 189]. Thus, for a given spatial continuity model, the kriging variance is not affected by the original data variance, and there may be equal kriging variances in situations where the data variance is completely different.

This article proposes a methodology for defining the optimal drilling grid in a lateritic nickel deposit based on the measurement of uncertainty, using the technique of conditional geostatistical simulation [5]. This method aims to reproduce the spatial continuity and intrinsic variability of the original data, and to combine multiple equally probable models in order to determine the associated uncertainty in the variables under study, thus enabling the appropriate sampling pattern for a given mineral deposit to be found.

## 2  Sequential Gaussian Simulation

The first sequential simulation methodology used in this study is sequential Gaussian simulation (SGS). This is an extension of the sequential conditional simulation algorithm, and is based on a Gaussian random function model. According to Pilger [6], the SGS method proposed by Isaaks [7] is characterised by the application of a sequential algorithm to the local univariate conditional distributions resulting from the decomposition of a particular Gaussian multivariate probability density function, controlled by a Gaussian multivariate model characterised by a covariance function $C(h)$.

In the SGS method, the local conditional probability distribution is determined using simple kriging, which defines the mean and variance of the distribution. This method assumes that the distribution is stationary and follows the form of a normal distribution, that is, with a mean of zero and variance of one. The element of interest (Ni in this case) rarely has values following this type of distribution, with the most common being asymmetric distributions with a few extreme values (positive asymmetry). It is therefore necessary to transform the distribution of the original data to a normal distribution [4] to enable sequential Gaussian simulation to be used. A process called data normalisation is used to assign a corresponding value to each original data item in the normal space. After SGS has been applied, it is then necessary to re-transform the normal values to their corresponding values in the original space.

Normalisation of the distribution of the original data is carried out using a transformation based on an increasing monotonic function. This function is called Gaussian anamorphosis [8], and can be written as follows:

$$z = f(\Phi)$$

where:
z = original data;
$\phi$ = transformation function;
y = normalised data for the respective value of z.

As described by Rivoirard [8], for each value of Z, the corresponding value in normal space is obtained from the accumulated distribution function of Z values (F(z)) and the accumulated distribution function of a standard normal variable Y(G(y)). Figure 1 shows this transformation. In addition, the mathematical translation of this methodology can be expressed as follows:

$$y = G^{-1}(F(z))$$

where:
y = normalised data for the respective value of Z;
G(y) = accumulated distribution function of a standard normal variable y;
F(z) = accumulated distribution function of Z values;
$G^{-1}(F(z))$ = standard normalised value whose cumulative probability is equal to F(z).

The reverse transformation process consists of transforming the values obtained from SGS in the normal space into their respective values in the original space. This reverse transformation can be performed by the inverse of normalisation; that is, for each value in the normal space (y), a value in the original space (z) is assigned that has the same accumulated probability (F(z) = G(y)).

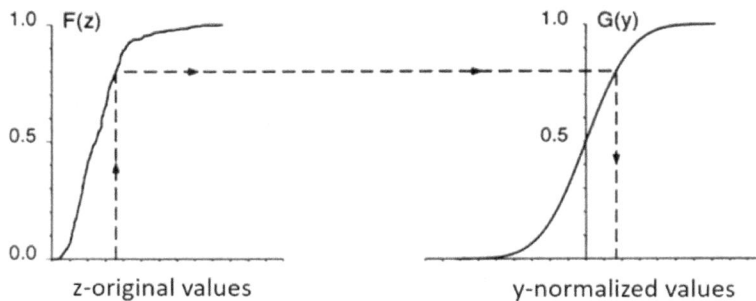

**Fig. 1** Example of transformation of an original distribution into a normalised equivalent (modified from [4])

# 3 Sequential Indicator Simulation

The second methodology used in this study is sequential indicator simulation (SIS) [9]. SSI is widely used to model categorical variables, and is based on a nonlinear kriging algorithm called indicator kriging [10]. Unlike SGS, this simulation algorithm is classified as nonparametric. Models of categorical variables are commonly created to represent weathering profiles or rock types, and it is during the modelling stage that important decisions are made such as the definition of the volume to be estimated and the stationarity of the data within the modelled domains for the different categorical variables. The models of these categorical variables are usually constructed explicitly, and depend on the interpretation and judgment of the specialist responsible for modelling the deposit. In many cases, there is insufficient data to allow for reliable (explicit) deterministic modelling, and a stochastic modelling algorithm is therefore used to build multiple realisations.

There are some drawbacks associated with SIS; for example, the variograms of the indicators control only the spatial relationship between two points in the model. SIS can also lead to geologically unrealistic or incorrect transitions between the simulated categories, and the cross-correlation between multiple categories is not explicitly controlled.

Despite these criticisms, there are many good reasons to consider SIS, such as the fact that the necessary statistical parameters are easy to infer from limited data. In addition, the algorithm is robust and provides a simple way to transfer uncertainty into categories using the resulting numerical models.

According to Journel [10], an indicator can be defined as follows:

$$i(u, Z_k) = \begin{cases} 1, & \text{for } z(u) \leq Z \\ 0, & \text{other cases} \end{cases} \quad K = 1, \ldots, K$$

This model uses K different categories that are mutually exclusive and exhaustive; that is, only one category can exist in a particular location.

# 4 Optimisation of a Drilling Grid

Geostatistical simulation is an excellent tool for assessing the uncertainty in the values of a given attribute or the probability of these values being above a given limit. This is because geostatistical simulation algorithms allow several possible scenarios to be constructed for the distribution of attribute values, i.e. the distribution of possible values at each simulated grid node.

Using geostatistical simulation, Pilger [11] quantified the uncertainty at each location using different uncertainty indices. In this way, each additional drillhole was located according to the values of the uncertainty indices. Pilger et al. [12] added sampling in regions of high uncertainty, after the insertion of each additional piece of information, a new simulation was performed, and the uncertainty indexes were recalculated.

The reduction of uncertainty to a theoretical and operational limit was observed after the addition of each piece of information related to the resources of a coal deposit. According to the authors, the theoretical limit is that above which the addition of new drilling sites becomes ineffective in terms of reducing uncertainty, and the operational limit is related to the cost of an extra drill hole.

Koppe [13] used geostatistical simulation to analyse the efficiency of different configurations of additional samples in terms of reducing uncertainty about a function and the factors that influence this efficiency. In her thesis, Koppe [13] presented the algorithm for the automatic construction of sample configurations and the computational workflow created to speed up the approximation of the uncertainty value on the transfer function obtained for each tested configuration.

# 5 Case Study

## 5.1 Methodology

The geometry of the ore body is defined by a non-continuous axis, approximately 18 km along its main direction (N12° E), and 0.9 to 2.4 km perpendicular to this direction (N102° E). The thickness varies from 0.5 to 34 m, with an average of 3.8 m. Due to this geometry, in which the horizontal dimensions are significantly larger than the vertical, the ore body was treated as being 2D.

The technique involves generating 100 simulations of the grade of the nickel, the thickness and the type of ore at a 90% confidence level, i.e. taking values between the 5th and 95th percentiles to measure the dispersion over the average, and hence finding the error for the block with a volume equivalent to three months of production.

## 5.2 Geostatistical Simulation with Original Database

Using the original drilling database distribution, 100 simulations were performed for the variables of ore type, thickness and nickel grades. For the ore type indicator simulation, the gslib program *blocksis* was used, whereas for the other two variables, the *sgsim* program was used.

The results from the 100 ore type simulations were used to condition the thickness and nickel grade simulations, so that the thickness and nickel grade were only simulated in cases where the presence of ore was predicted by the ore type indicator simulation. The simulations were carried out using grid of dimensions 5 × 5 × 1 m, and were divided into five files due to space limitations.

Figure 2 shows the parameter files for the *blocksis* and *sgsim* programs.

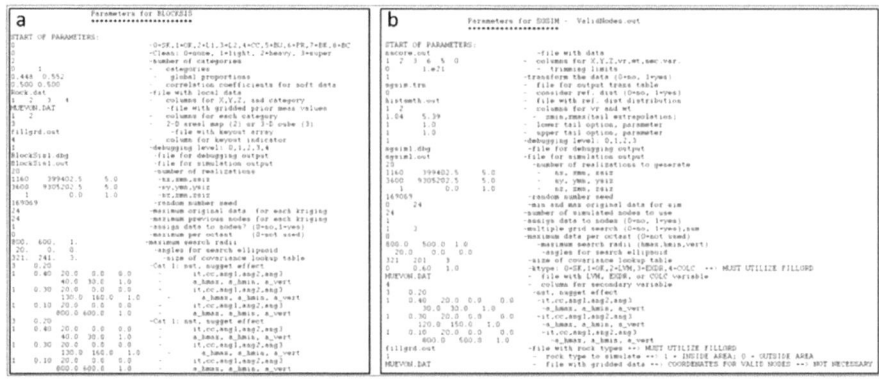

**Fig. 2** Parameter files for **a** *blocksis* and **b** *sgsim* programs

In order to guarantee the adherence of these simulations to the original data, it was necessary to validate the variograms, histograms and averages. Figure 3 shows the validation of the variograms for 100 simulations of the nickel grade, thickness and ore type variables.

The results of the simulation for the nickel grade and thickness over a grid of 5 × 5 × 1 m were reblocked to a scale of 350 × 350 × 1 m, representing the mass over three months of production. For this, we used the program *modelrescale* from GSLib, and at this scale, the uncertainties in the nickel grade and thickness variables were assessed using the rule of 15% of error with 90% confidence.

To obtain the results in terms of the metal content, the results for the nickel and thickness were multiplied directly, since the area and density used in this study were treated as constant and the intention was to assess only the error in the metal content and not its absolute value.

Figure 4 shows the errors for the metal content versus the number of drillholes within each panel for three months of production. The red horizontal line shows an error of 15% with 90% confidence, the blue cross represents the error in each panel for three months versus the number of drill holes, and the red squares show the average error for the panels for the same numbers of holes. The solid black line represents a logarithmic adjustment function.

It can be concluded that the current drilling distribution in the project is not sufficient to determine the optimal drilling grid, since the number of drillholes and the error shown by the blocks for three months of production are not sufficient to reach the ideal value of 15% error with 90% confidence (the red line in Fig. 4).

Hence, a virtual drilling grid was created with additional drillholes in order to test the methodology and to define an appropriate grid.

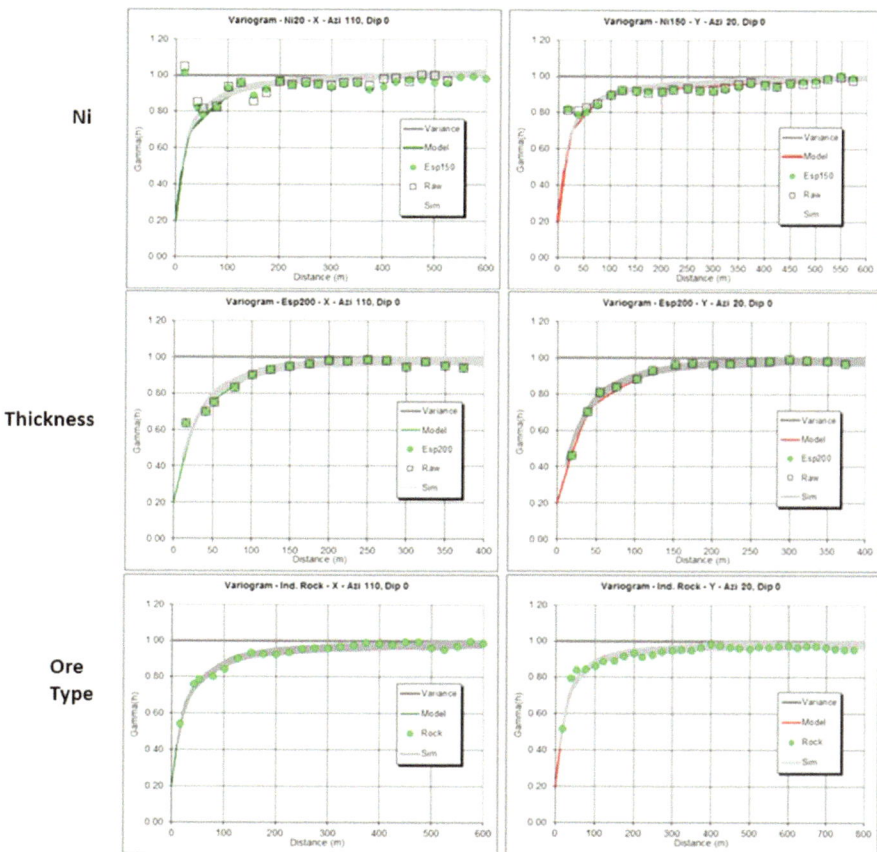

**Fig. 3** Validation of variograms for 100 simulations of nickel grade, thickness and ore type

## 5.3   Geostatistical Simulation with a Virtual Drilling Grid Database

The virtual drilling grid used for this test was 25 × 25 m. To create this grid, the entire process described above was applied, although only one simulation involving the ore type, nickel and thickness was performed.

## 5.4   Geostatistical Simulation of 100 Realisations of Thickness, Nickel and Ore Type

Using the results derived for the ore type, thickness and nickel simulations in the previous step, a new database was created for these variables over a grid of 25 ×

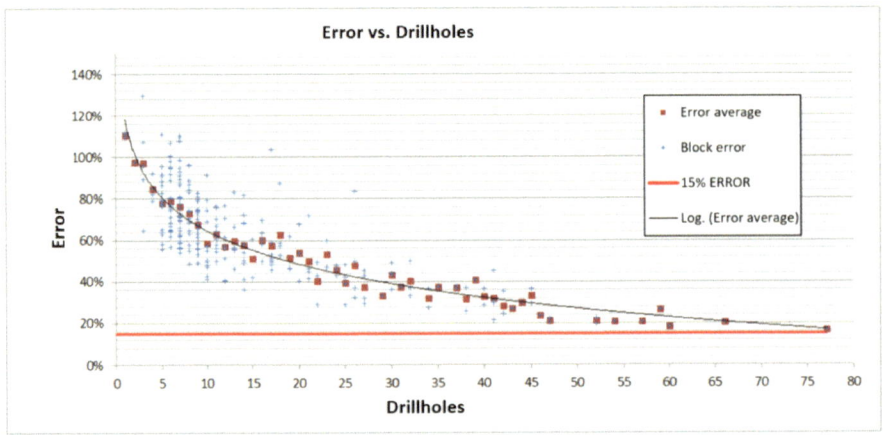

**Fig. 4** Error in the metal content versus the number of drillholes within the panels for 3 months of production. The red horizontal line shows an error of 15% with 90% confidence

25 m (a simulated virtual database). This database was then used in a new simulation based on a 5 × 5 m grid with 100 realisations.

As in all stages of this study, it was necessary to validate the simulations with the data for the 25 × 25 m, which is referred to in the following as the original grid. Figure 5 shows the validation of the histograms for the nickel grade and the thickness. Note the small range of variation between the simulation histograms. Figure 6 shows the validation of the averages and proportions, where the average for each realisation is calculated and compared with the average from the database.

Figure 7 shows the validation of the variograms. The 100 realisations variogram ergodic fluctuations for the three variables represent the data in the original 25 × 25 m grid along the main directions.

The grid scale of 5 × 5 × 1 m was reblocked to 350 × 350 × 1 m, using the *modelrescale* program from gslib.

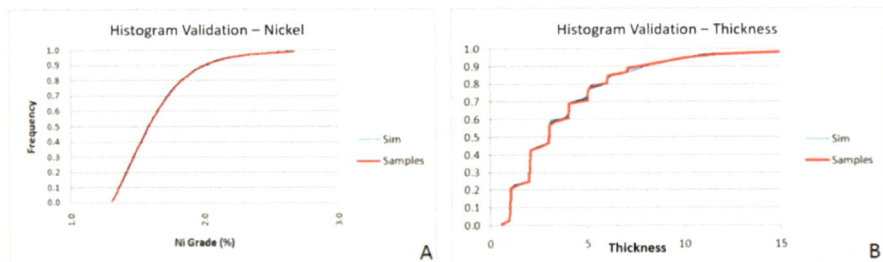

**Fig. 5** Histogram validation for **a** nickel grade and **b** thickness

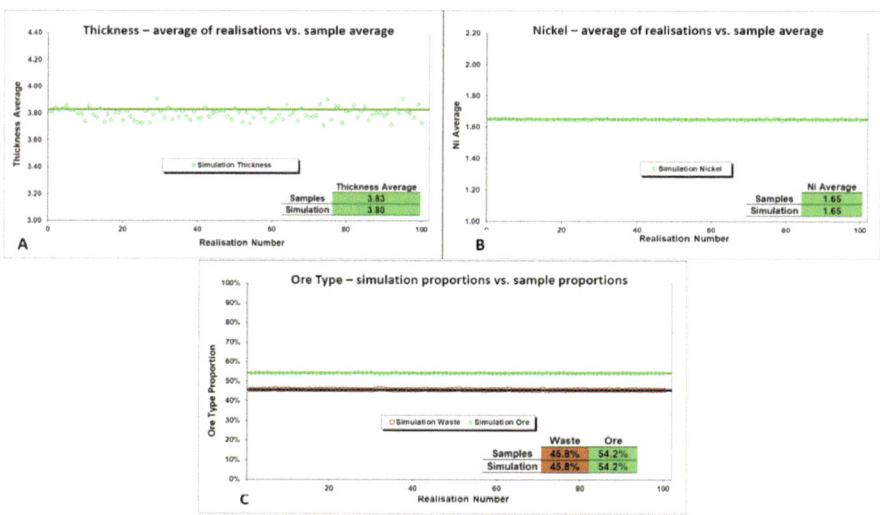

**Fig. 6** Validation of averages for **a** thickness, **b** nickel and **c** rock proportions

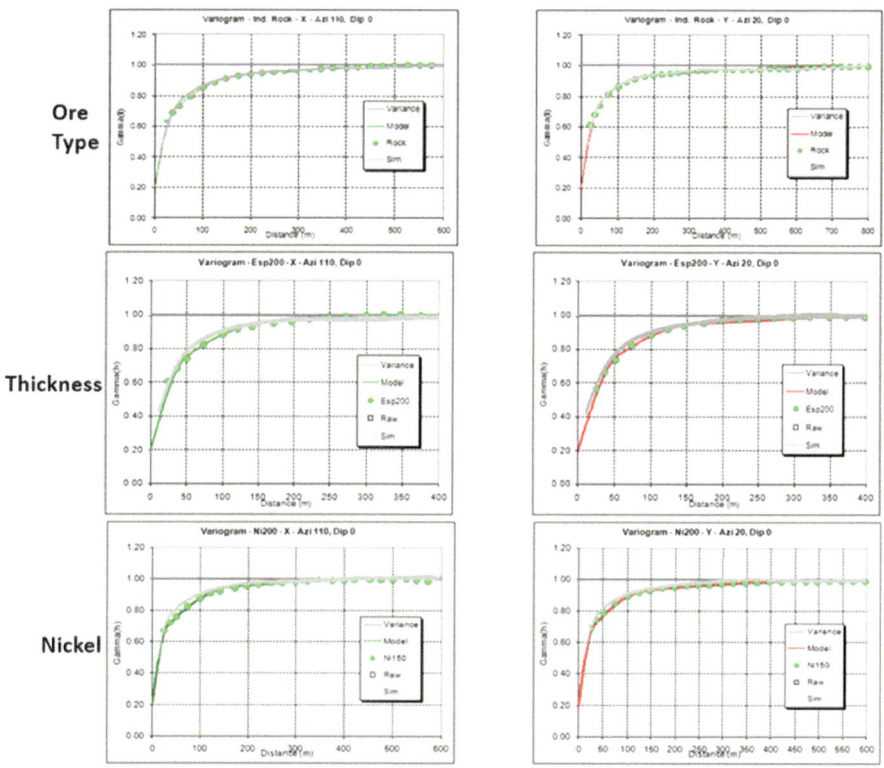

**Fig. 7** 100 realisations variogram ergodic fluctuations for ore type, thickness and nickel grade

# 6    Results and Discussion

Figure 8 shows the panels for three months of production with their respective errors, where the red blocks indicate an error of less than 15% at a confidence level of 90%.

Since panels with less than 50% chance of being ore may be defined as waste, only panels with a probability of greater than 50% of being ore were considered in this study. A graph of the relationship between the error and ore percentage is shown in Fig. 9.

Note that the error is reduced as the proportion of ore in the panel increases. This gives rise to the possibility of optimising the drilling grid in the regions with a higher probability of ore.

**Fig. 8**   Three months of production panels with a probability of being ore of higher than a 50%

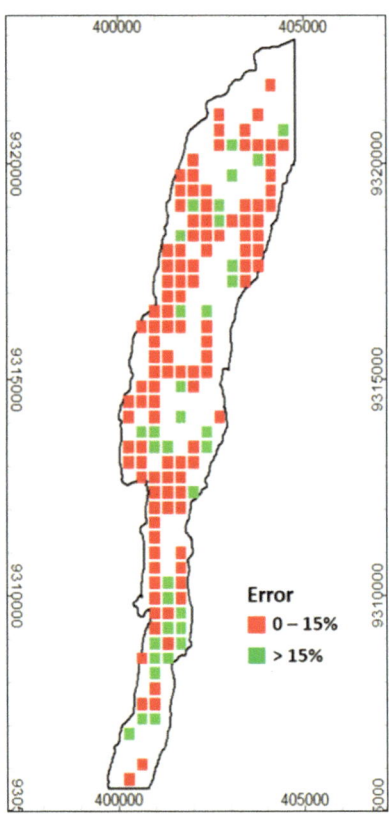

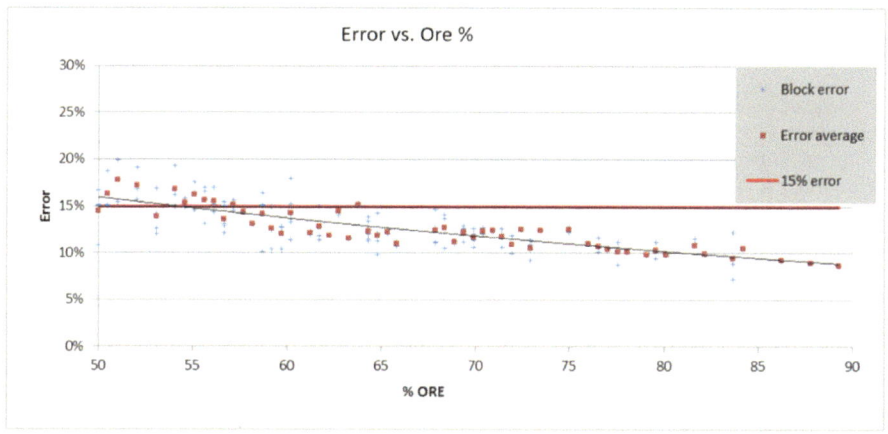

**Fig. 9** Error in metal content versus ore percentage for each panel. The red horizontal line shows 15% error with 90% confidence

## 7 Conclusions

The applicability of the proposed methodology was evaluated, and the results indicate that the $25 \times 25$ m grid meets the requirements of an error of less than 15% with 90% confidence in the panels selected as ore.

An important aspect of the methodology proposed in this study is that it does not depend on the existence of a geological model, since the existence or otherwise of a mineralised zone is defined using SIS. This technique can also therefore be recommended for applications involving mineral deposits where there is no detailed knowledge, i.e. during the intermediate stages of exploration.

**Acknowledgements** The authors would like to thank UFRGS and Anglo American for technical and financial support.

## References

1. Matheron, G.: Principles of geostatistics. Econ. Geol. **58**, 1246–1266 (1963)
2. David, M.: Geostatistical Ore Reserve Estimation. Developments in Geomathematics, vol. 2, 364p. Elsevier Scientific Publishing Company, Amsterdam (1977)
3. Isaaks, E.H., Srivastava, M.R.: An Introduction to Applied Geostatistics, 561p. Oxford University Press, New York (1989)
4. Goovaerts, P.: Geostatistics for Natural Resources Evaluation, 483p. Oxford University Press, New York (1997)
5. Journel, A.G.: Geostatistics for conditional simulation of ore bodies. Econ. Geol. **69**(5), 673–687 (1974)
6. Pilger, G.G.: Aumento da Eficiência dos Métodos Sequenciais de Simulação Condicional, 229p. Universidade Federal do Rio Grande do Sul, Tese de Doutorado (2005)

7. Isaaks, E.H.: The application of monte Carlo methods to the analysis of spatially correlated data, 213p. Ph.D. thesis, Stanford University, USA (1990)
8. Rivoirard, J.: Introduction to Disjunctive Kriging and Nonlinear Geostatistics, 89p. Centre de Géostatistique, Ecole des Mines de Paris (1990)
9. Alabert, F.G.: Stochastic imaging of spatial distributions using hard and soft information. M.Sc. thesis, Stanford University, California (1987)
10. Journel, A.G.: The indicator approach to estimation of spatial distributions. In: Proceedings of the 17th APCOM (International Symposium on the Application of Computers and Mathematics in the Mineral Industry), SME-AIME, Golden, Colorado, EUA, pp. 793–806 (1982)
11. Pilger, G.G.: Critérios para Locação Amostral Baseados em Simulação Estocástica. Dissertação de Mestrado, 127p. Programa de Pós-Graduação em Engenharia de Minas, Metalúrgica e de Materiais (PPGEM). Universidade Federal do Rio Grande do Sul (2000)
12. Pilger, G.G., Costa, J.F.C.L., Koppe, J.C.: Optimizing the value of a sample. In: Application of Computers and Operations Research in the Mineral Industry, Phoenix, Proceedings of the 30th International Symposium, vol. 1, pp. 85–94. Society for Mining, Metallurgy and Exploration, Inc. (SME), Littleton (2002)
13. Koppe, V.C.: Metodologia para Comparar a Eficiência de Alternativas para Disposição de Amostras, 236p. Tese de Doutorado. Programa de Pós-Graduação em Engenharia de Minas, Metalúrgica e de Materiais (PPGEM). Universidade Federal do Rio Grande do Sul (2009)

# LSTM-Based Deep Learning Method for Automated Detection of Geophysical Signatures in Mining

Mehala Balamurali and Katherine L. Silversides

**Abstract** The mining of stratified ore deposits requires detailed knowledge of the location of orebody boundaries. In the Banded Iron Formation (BIF) hosted iron ore deposits located in the Pilbara region of Western Australia the natural gamma logs are useful tool to identify stratigraphic boundaries. However, manually interpreting these features is subjective and time consuming due to the large volume of data. In this study, we propose a novel approach to automatically detect natural gamma signatures. We implemented a LSTM based algorithm for automated detection of signatures. We achieved a relatively high accuracy using gamma sequences with and without added noise. Further, no feature extraction or selection is performed in this work. Hence, LSTM can be used to detect different signatures in natural gamma logs even with noise. So, this system can be introduced in mining as an aid for geoscientists.

**Keywords** Long short-term memory · Deep learning · Natural gamma signals

## 1 Introduction

In the Banded Iron Formation (BIF) hosted iron ore deposits located in the Pilbara region of Western Australia, natural gamma logs are frequently used to determine the location of stratigraphic boundaries [1]. The location of these boundaries is required for accurate modelling of the deposit. These deposits consist of layers of BIF and shale, with sections of the BIF enriched to create a minable high grade ore [2]. This paper studies a typical Marra Mamba type iron ore deposit, with iron ore in the Mount Newman Member overlain by the shale in the West Angelas Member.

The natural gamma logs are one of several geophysical logs that are typically collected in exploration holes, along with density and magnetic susceptibility. As

M. Balamurali (✉) · K. L. Silversides
Australian Centre for Field Robotic, University of Sydney, Camperdown, Australia
e-mail: mehala.balamurali@sydney.edu.au

K. L. Silversides
e-mail: katherine.silversides@sydney.edu.au

© The Author(s) 2023 163
S. A. Avalos Sotomayor et al. (eds.), *Geostatistics Toronto 2021*, Springer Proceedings in Earth and Environmental Sciences, https://doi.org/10.1007/978-3-031-19845-8_14

thousands of holes may be drilled for a single deposit, manually interpreting them can be an arduous process. Additionally, it is prone to inconsistencies between different interpreters as well as errors. Therefore an automatic method that can assist the interpretation is greatly desired. A computer-aided method is a better option as it can provide a fast, objective, and reliable analysis. Many studies have been conducted on the development of automatically identified boundaries based on machine learning applications, including several studies on these types of deposit [3, 4]. However, the process of choosing a set of optimal features to classify a signature from the signal is very difficult. Therefore, a deep learning technique is presented in this study to overcome the challenges faced by conventional automated systems. A deep learning network involves several stages of learning processes, including an input layer, hidden layers, and an output layer [5]. The network takes unprocessed data as the input and learns the representative features that needed for classification without user input. The network is trained using the backpropagation algorithm.

In this study we investigated the capability of a long short-term memory (LSTM) network to classify several geological signatures from a gamma log. LSTM is a type of recurrent neural network mostly used to analyse time series sequence data [6]. In this study the authors investigated the capability of a LSTM network to classify several geological signatures from a gamma log sequence data along the down hole samples. The work demonstrates a classifier with three different outputs that can differentiate AS1-AS2 signals and NS3-NS4 signals from long gamma sequence.

## 2   Data Used

Natural gamma logs from a typical Marra Mamba style iron ore deposit were used. This deposit contains two natural gamma signatures of particular interest. The first signature is produced by the AS1-AS2 shales at the base of the West Angelas Member (Fig. 1a) and marks the transition between the shale and the Mount Newman ore below. The second signature comes from the NS3-NS4 shale bands (Fig. 1b) that mark a transition between two geological units within the Mount Newman Member. 42 examples of each of these signatures were chosen, along with 42 examples of other signatures from different parts of the natural gamma logs.

## 3   Methodology

### 3.1   Long Short-Term Memory (LSTM)

In order to train a deep neural network to classify each signal along the depth of sequence data, a sequence-to-sequence LSTM network was used as proposed in Matlab R2019a [7]. The existing method was used to make prediction for each time

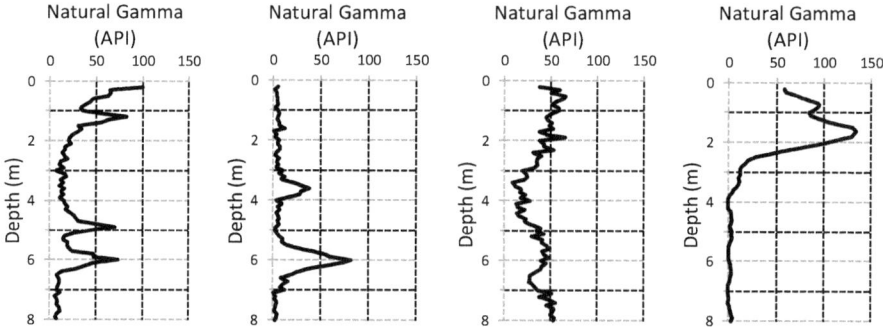

**Fig. 1** Typical examples of the natural gamma signatures for **a** AS1-AS2, and **b** NS3-NS4. **c, d** Contain two examples of other signatures, there is a large variety of signatures in this group

step of the sequence. However, in this study, the method was used to make prediction of depth sequence data.

The proposed method consists of a LSTM layer, with tanh as the activation function. LSTM layer was set to have 200 hidden units and output the full sequence. The final layer, which is a fully connected (FC) layer with a softmax activation function, is used as the classifier with three classes. The classifier uses the output from the LSTM as the input and predicts the class label of the gamma signatures for a given sequence. The 'adam' optimizer was used and training was happened with different number of epochs.

## 3.2 Training and Validation

In order to analyze the capability of the LSTM architecture for identifying gamma signatures, a range of tests were performed. Out of 42 sequences, 30 sequences were used as training samples and rest were used as test samples. Two training libraries were used. In the first library all samples had the same signal order. The other library had two different sequence sets: AS1-AS2, other and NS3-NS4 in order, NS3-NS4, AS1-AS2 and other in order.

Figure 2 shows twelve test samples each containing a combination of AS1-AS2 and NS3-NS4 signatures along with signatures from other sections of the logs. The test sequences have the AS1-AS2, NS3-NS4 and other signals in different order. The first six test sequences have the signals in the order AS1-AS2, other and NS3-NS4, and the other six sequences have the order NS3-NS4, AS1-AS2 and other. To test the robustness of this method to noise, the test was re-run with random noise added to the test samples.

In total four tests were completed. Case 1 and Case 2 were trained using libraries 1 and 2 respectively and were both stopped at epoch 200. Case 3 was trained with

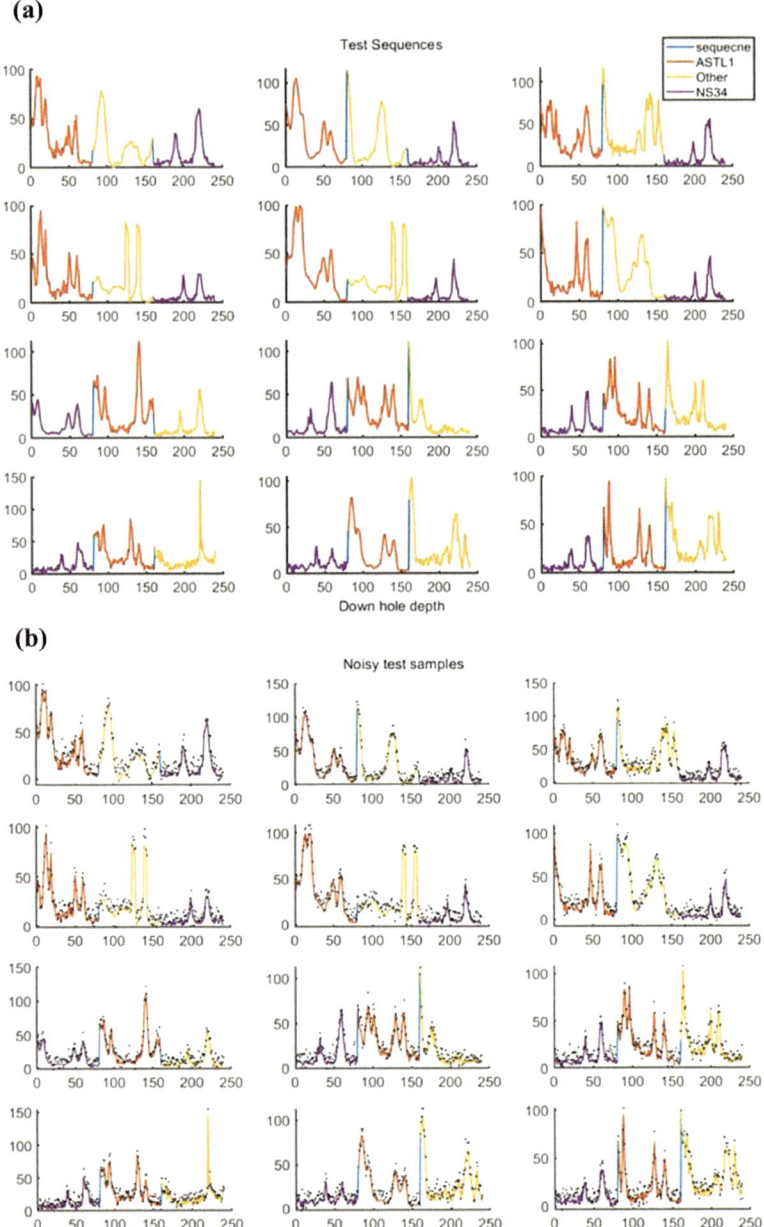

**Fig. 2** Test sequences with the combination of AS1-AS2 (orange), NS3-NS4 (purple) signatures and other (yellow). Samples with (**a**) no noise and (**b**) noise added. Blue lines show the joins between the different signatures

library 2 and stopped at epoch 228. Case 4 used the same training as Case 3, but predicted the signature classes of samples with added noise.

## 4   Results and Discussion

The accuracy and loss of the proposed method from the training as a function of epochs can be seen in Fig. 3. Each plot is associated with different training library and different epochs. In case one library 1 was used to train the net. Then the net was used to predict the signatures in the original test samples (no noise added) (Fig. 2a). In other cases, library 2 was used in the training process with epochs 200 and 228. Prediction accuracy was tested on both original and noisy samples (Fig. 2). In all scenarios a high accuracy were achieved during the training process (Fig. 3). The performance is summarized in Figs. 4 and 5.

In Fig. 4a it can be seen that the classification accuracies are significantly higher in the first 6 sequences compared to the remaining six sequences. This is because the net used in prediction was trained with library 1, which has all signals in the same order. Without the inclusion of different signal orders in the training process, the LSTM method failed to identify signals where the order was different. In comparison, when training with mixed order samples the LSTM closely predicted the signals in 11 out of 12 samples (Fig. 4b). The accuracy of the prediction was further improved at epoch 228, where the author stopped the training where accuracy of the network peaked (Fig. 4c). The same trained network obtained relatively good accuracy in the noisy test samples as well (Fig. 4d.). In all scenarios test sample 7 yielded a much lower accuracy. This is because the AS1-AS2 signature in that sample is quite different to the other AS1-AS2 signatures.

## 5   Conclusion

In this paper, a deep learning based solution using LSTM is developed for natural gamma signature identification. The proposed method learns patterns from gamma sequences, and could successfully identify the significant AS1-AS2 and NS3-NS4 signatures. With only a few training samples it generally achieved a relatively high accuracy. However, the accuracy was dependent on the signal order present in the training samples. Further evaluation is needed with increasing training samples with different sequence lengths.

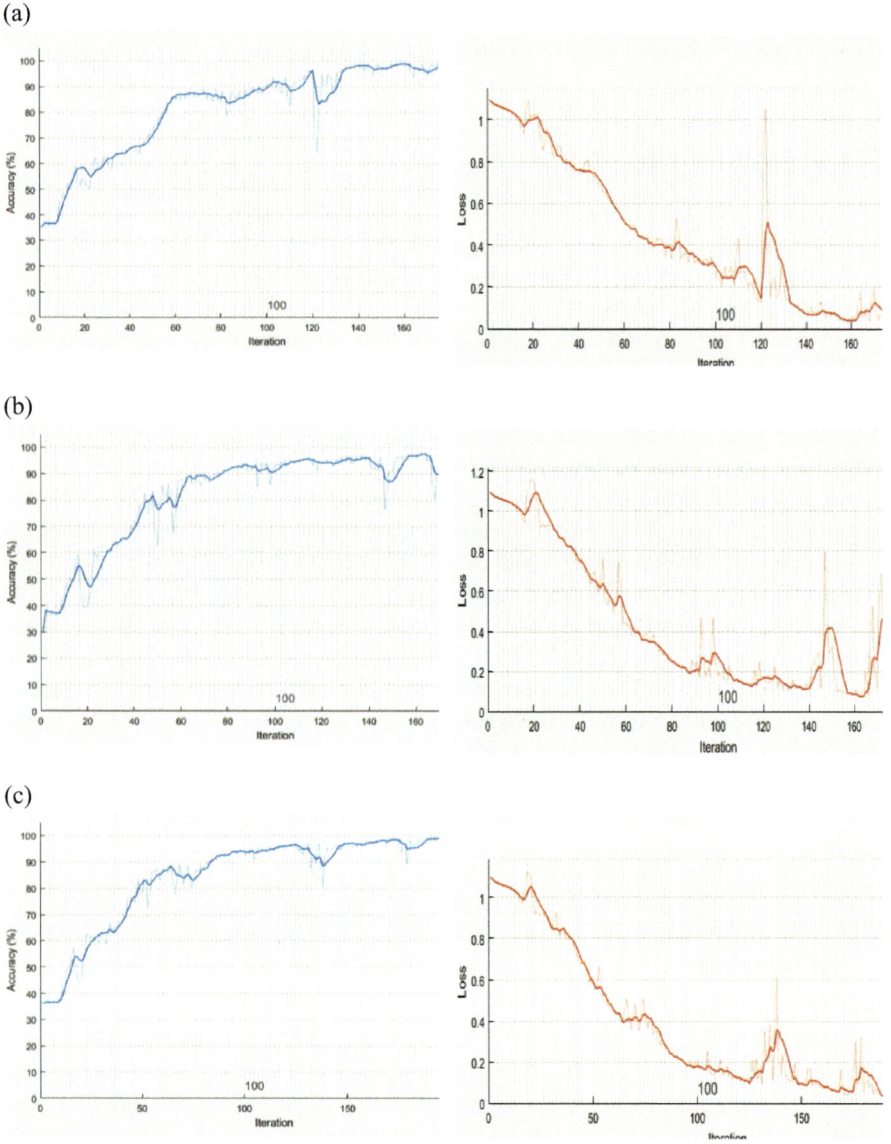

**Fig. 3** Performance from the proposed LSTM method. Left: accuracy versus epochs; Right: Loss versus epochs; **a** Training with library 1: AS1-AS2, Other, NS3-NS4, training accuracy at epoch 200. **b** Training with library 2: order 1: AS1-AS2, Other,NS3-NS4; order2: NS3-NS4, AS1-AS2, other at epoch 200. **c** Training with library 2 manually terminated at epoch 228

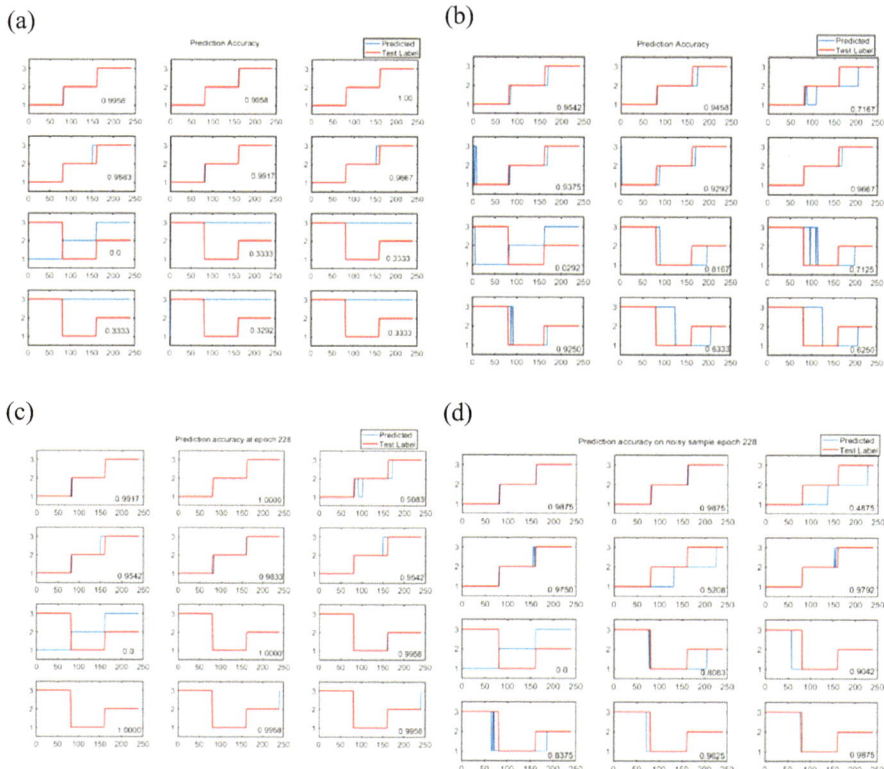

**Fig. 4** Comparing predicted class labels (blue) with given class labels (red) for different scenarios: **a** Training with library 1, epoch 200, **b** Training with library 2, epoch 200, **c** Training with library 2, epoch 228, **d** Training with library 2, epoch 228 and prediction on noisy samples. X axis: depth; Y axis: class labels, 1:3 represent the AS1-AS2 signatures, other signatures and NS3-NS4 signatures respectively

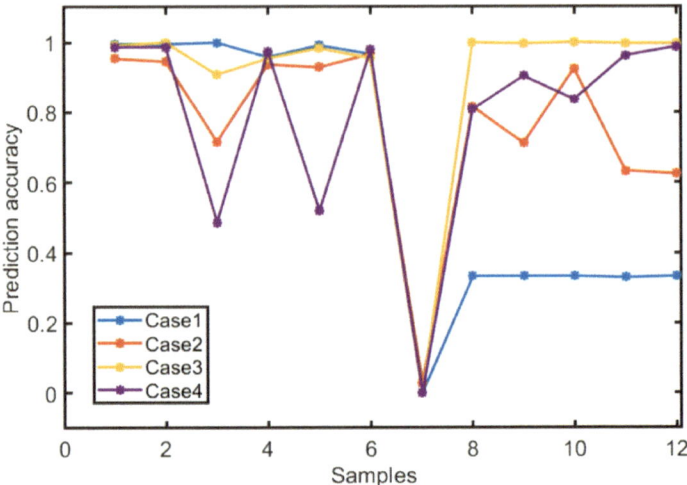

**Fig. 5** Summaries of the prediction accuracy on 12 samples for each case. Case1: Training with library 1, epoch 200, Case 2 Training with library 2, epoch 200, Case 3 Training with library 2, epoch 228, Case 4 Training with library 2, epoch 228 and prediction on noisy sample

**Acknowledgements** This work has been supported by the Australian Centre for Field Robotics and the Rio Tinto Centre for Mine Automation.

# References

1. Harmsworth, R.A., Kneeshaw, M., Morris, R.C., Robinson, C.J., Shrivastava, P.K.: BIF-derived iron ores of the Hamersley Province. In: Hughes, F.E. (eds.) Geology of the Mineral Deposits of Australia and Papua New Guinea, pp. 617–642. The Australasian Institution of Mining and Metallurgy, Melbourne (1990)
2. Thorne, S.W., Hagemann, S., Webb, A., Clout, J.: Banded iron formation-related iron ore deposits of the Hamersley Province, Western Australia. In: Hagemann, S., Rosiere, C., Gutzmer, J., Beukes, N.J. (eds.) Banded Iron Formation-Related High Grade Iron Ore; Rev. Econ. Geol. **15**, 197–221 (2008)
3. Silversides, K., Melkumyan, M., Wyman, D., Hatherly, P.: Automated recognition of stratigraphic marker shales from geophysical logs in iron ore deposits. Comp. Geo. **77**, 118–125 (2015). https://doi.org/10.1016/j.cageo.2015.02.002
4. Nathan, D., Duuring, P., Holden, E.J., Wedge, D., Horrocks, T.: Learning characteristic natural gamma shale marker signatures in iron ore deposits. Comp. Geo. **106**, 77–88 (2017). https://doi.org/10.1016/j.cageo.2017.06.001
5. LeCun, Y., Bengio, Y., Hinton, G.: Deep learning. Nature **521**, 436–444 (2015)
6. Hochreiter, S., Schmidhuber, J.: Long short-term memory. Neural Comput. **9**(8), 1735–1780 (1997)
7. MathWorks: Matlab R2019a documentation. https://au.mathworks.com/help/deeplearning/examples/sequence-to-sequence-classification-using-deep-learning.html. Last accessed 01 May 2019

# Earth Science

# RETRACTED CHAPTER:
# Spatio-Temporal Optimization
# of Groundwater Monitoring Network
# at Pickering Nuclear Generating Station

**Pedram Masoudi, Yvon Desnoyers, and Mike Grey**

The authors have retracted this Conference Paper because the data owner did not approve the publication. The content of this Conference Paper has been removed for that reason. All authors agree with this retraction.

P. Masoudi (✉) · Y. Desnoyers
Geovariances, 77210 Avon, France
e-mail: masoudi@geovariances.com

M. Grey
Kinectrics, Toronto, ON, Canada

© The Author(s) 2023, corrected publication 2023
S. A. Avalos Sotomayor et al. (eds.), *Geostatistics Toronto 2021*, Springer Proceedings
in Earth and Environmental Sciences, https://doi.org/10.1007/978-3-031-19845-8_15

# Domains

# Applying Clustering Techniques and Geostatistics to the Definition of Domains for Modelling

Gabriel de Castro Moreira, João Felipe Coimbra Leite Costa, and Diego Machado Marques

**Abstract** Machine learning is a broad field of study that can be applied in many areas of science. In mining, it has already been used in many cases, for example, in the mineral sorting process, in resource modeling, and for the prediction of metallurgical variables. In this paper, we use for defining estimation domains, which is one of the first and most important steps to be taken in the entire modeling process. In unsupervised learning, cluster analysis can provide some interesting solutions for dealing with the stationarity in defining domains. However, choosing the most appropriate technique and validating the results can be challenging when performing cluster analysis because there are no predefined labels for reference. Several methods must be used simultaneously to make the conclusions more reliable. When applying cluster analysis to the modeling of mineral resources, geological information is crucial and must also be used to validate the results. Mining is a dynamic activity, and new information is constantly added to the database. Repeating the whole clustering process each time new samples are collected would be impractical, so we propose using supervised learning algorithms for the automatic classification of new samples. As an illustration, a dataset from a phosphate and titanium deposit is used to demonstrate the proposed workflow. Automating methods and procedures can significantly increase the reproducibility of the modeling process, an essential condition in evaluating mineral resources, especially for auditing purposes. However, although very effective in the decision-making process, the methods herein presented are not yet fully automated, requiring prior knowledge and good judgment.

G. de Castro Moreira (✉)
Department of Mining Engineering, Universidade Federal do Rio Grande do Sul—UFRGS, Av. Bento Gonçalves, 9500, Setor 4, Predio 75, Sala 101, Porto Alegre, Rio Grande Do Sul, Brazil
e-mail: gabrielcm.moreira@gmail.com

J. F. C. L. Costa
Department of Mining Engineering, Universidade Federal Do Rio Grande Do Sul—UFRGS, Porto Alegre, Rio Grande Do Sul, Brazil
e-mail: jfelipe@ufrgs.br

D. M. Marques
Department of Geology (Geosciences Institute—IGEO), Universidade Federal Do Rio Grande Do Sul—UFRGS, Porto Alegre, Rio Grande Do Sul, Brazil
e-mail: diego.marques@ufrgs.br

© The Author(s) 2023
S. A. Avalos Sotomayor et al. (eds.), *Geostatistics Toronto 2021*, Springer Proceedings in Earth and Environmental Sciences, https://doi.org/10.1007/978-3-031-19845-8_16

**Keywords** Machine learning · Cluster analysis · Estimation domains · Mining

# 1  Introduction

## 1.1  Machine Learning in Mining

Machine learning (ML), a term introduced by [31], enables computers to learn from data to perform tasks then and make decisions based on the mathematic models that were built, without giving specific commands to these computers. It has been heavily applied in many fields of science and industry. Although not a new subject, it has enjoyed high popularity and a sharp increase in development over the last decades, thanks to the growth of computer power and easy access to the public.

ML is a complex and broad scientific field and must be used with caution. It can offer interesting solutions, especially regarding complex problems with big databases and high dimensionality, when properly applied. In the mining industry, it has already been applied in several tasks, for example in the definition of domains for resource modeling (e.g., [12, 18, 19, 27]), in the mineral analysis and sorting process (e.g., [6, 14, 33]) and for the prediction of metallurgical variables (e.g., [16, 22, 23]).

This paper specifically addresses the matter of defining domains for modeling using cluster analysis. We compare different algorithms and elaborate on the formal validation of the spatial distribution of the resulting clusters using correlograms of the indicators, as suggested in [21].

Additionally, we discuss and apply supervised learning algorithms for the automatic classification of new samples. Consequently, incorporating new data into the database respects the same rules used to define those domains, which has been suggested by [27] and is now applied in a real case scenario with a phosphate-titanium illustration case.

## 1.2  Stationarity in the Context of Mineral Resource Modeling

The concept of stationarity is closely related to the homogeneity of geological bodies. We will assume that a phenomenon is stationary when it shows constant expected values, covariance moments, and autocorrelation structures across the study area in any given location, a simplified definition based on [15]. However, these characteristics are rarely present over large volumes of in situ natural materials. Therefore, it becomes necessary to distinguish more homogeneous portions inside mineral deposits, the so-called "modeling domains". Hence, the geologic block models are more accurate to the reality that they intend to represent.

Proper data statistics and a good understanding of the geological context are essential for an adequate segmentation of a mineral deposit into domains for modeling. Unsupervised machine learning techniques, specifically cluster analysis algorithms, can be especially suited for this matter. These methods can divide the available data into groups based on similarities and dissimilarities.

## 1.3 Types of Clustering Algorithms and Background

Traditionally, cluster analysis uses the relationships that are present in the data based only on statistical parameters, that is, the relationships in the attribute (multivariate) space (e.g., [7, 17, 34]), without considering the spatial connectivity or the geological aspects [29]. Thus, using such traditional algorithms on geostatistical datasets encounters considerable limitations, as statistical similarity does not necessarily imply geographic continuity.

In the last decades, new methods were introduced to analyze the clustering of spatial data. These algorithms offer the possibility of grouping data into spatially contiguous clusters and, at the same time, respect the statistical relationships between the variables (e.g., [3, 8, 18, 24, 27, 32]).

The method presented by Oliver and Webster [24] uses spatial analysis to determine the scale of spatial variation. This is then applied for clustering the samples into spatial contiguous groups. Ambroise et al. [1] introduced a technique where spatial constraints are applied to the expectation–maximization algorithm [5]. Scrucca [32] presented a method based on the autocorrelation statistics developed by Getis and Ord [13] and Ord and Getis [25].

Romary et al. [28] applied spatial clustering based on the traditional agglomerative hierarchical method and, later, Romary et al. [27] added a new approach, inspired on the spectral clustering algorithm. Fouedjio [8, 9] applied a non-parametric kernel estimator, developing an algorithm also based on the spectral approach. Fouedjio [10] then extended the spectral clustering approach presented by Fouedjio [8] to be applied to large scale geostatistical datasets.

Martin and Boisvert [18] introduced a method that involves: (i) introduction of a new algorithm based on clustering ensembles and (ii) a metric that combines the spatial and the multivariate character of the data.

D'Urso and Vitale [3] proposed a modification of the technique that was presented by Fouedjio [9], aiming at neutralizing eventual spatial outliers. [11] reviewed many of the spatial clustering techniques, categorizing them according to their characteristics.

## 1.4 Discussions on the Validation Process

All unsupervised techniques aim at finding patterns in unlabeled data, allowing the definition of groups based on their similarities/dissimilarities. This fact is what brings up one of the most significant challenges in cluster analysis: validation. Because there are no true values for reference, validation is a subjective task.

Some specific validation techniques, such as the Silhouette [30], the Davies-Bouldin [4], and the Calinski-Harabaz [2] methods can aid in this task, but only access the effectiveness of the clustering process in the multivariate space. Martin and Boisvert [18] proposed an alternative approach for measuring the clustering quality in the multivariate and geographic spaces simultaneously: the "dual-space metrics".

An additional challenge is to choose the most appropriate number of clusters. An excessive number of domains can unnecessarily complicate the subsequent steps in the modeling workflow (e.g., contour modeling, estimation, simulation). On the other hand, very few domains can imply the mixing of statistical populations, compromising the accuracy of the resulting models.

Therefore, there is no ultimate methodology for cluster analysis when applied to geostatistical datasets. The results have to be tested and different scenarios compared in both multivariate and geographic spaces. Nevertheless, what is being tested is not the veracity of the clusters, but their relative quality, their practical sense, in other words, whether or not it was possible to group the data satisfactorily.

## 1.5 Supervised Learning Applied to the Classification of New Samples

Mining is a dynamic activity, and new information is constantly added to the database in every operation. Performing the entire cluster analysis each time that new samples are collected would be somewhat impractical. Furthermore, as the clustering techniques are based on the search for complex relationships in the data, other configurations could arise, slightly different from those already defined, which would require the revision of the whole modeling process, including the definition of contours and the analysis of the spatial continuity (i.e. variography).

At the same time, the new samples must be labeled according to the previously conducted cluster analysis [27]. In other words, they are classified according to the same rules so that a new sample, when assigned to a particular group, is more similar to the other samples of the same cluster than to samples from other clusters. As these designations do not follow simple rules, classifying new samples is not a trivial task, and supervised machine learning techniques are especially suitable for this matter. This can be achieved with algorithms such as decision trees, random forests, support vector machines, and artificial neural networks.

Figure 1 presents a flowchart with the proposed steps for the integrated cluster analysis and automatic classification of samples in a mineral resource modeling context [20]. Sporadically, as the database grows in number, the definition of groups can be updated, as indicated by the dashed line in the flowchart of Fig. 1, using samples that had not been previously used in the cluster analysis. The supervised classifier must then be updated so that it incorporates the new information to be used in the classification of new samples in a continuous process.

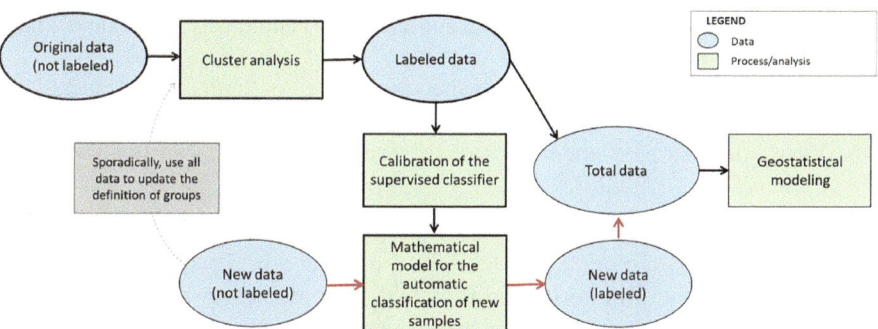

**Fig. 1** Proposed workflow for the integrated use of supervised and unsupervised machine learning methods in defining domains for modeling and classifying new samples in a mineral resource modeling process

## 2   Methods and Workflow

### 2.1   Clustering Algorithms

In order to observe the behavior of different types of clustering approaches, four algorithms were applied in this study. Two of them do not consider the spatial distribution of the samples. The other two are of the spatial type, which account for the relationships of data in the attribute space as well as for the position of the samples in the geographic space:

(i)   k-means [17]—one of the most traditional and widely used clustering algorithms;

(ii)   Agglomerative hierarchical [34]—another widely used technique, firstly introduced in the field of taxonomy, and later adopted in many fields of science;

(iii)   Dual-space clusterer [18]—a specially designed method to deal with spatial data, herein referred to as *dsclus*,

(iv)   Autocorrelation-based clusterer [32]—another algorithm for handling spatial data, based on the clustering of local autocorrelation statistics, herein mentioned as *acclus*.

### 2.2   Validation Methods

The following validation methods were applied to evaluate the results and help to choose the best clustering algorithm and configuration:

(i)   The Silhouette [30], the Calinski-Harabasz [2], and the Davies-Bouldin [4] scores;

(ii)   The dual space metrics [18], which comprehends spatial entropy (H) and within clusters sum of squares (wcss);

(iii)  Indicators correlograms to assess the geographic continuity of the clusters (as suggested in [21]);

(iv)   Visual inspection of the spatial distribution and statistical evaluation of the clusters.

The metrics indicated on item (i) measure how well the elements are clustered inside the respective groups using Euclidean distances in the multivariate space. Higher values of both Silhouette and Calinski-Harabasz scores and lower Davies-Bouldin scores are desired, indicating more organized clusters.

Martin and Boisvert [18] proposed the dual-space metrics for simultaneously verifying multivariate and spatial cohesion of clusters. H and wcss are calculated independently but simultaneously evaluated, so the goodness of the clusterings are evaluated in geographic and multivariate spaces.

Generally, the best configurations for spatial data clustering are not those with the lowest multivariate scores nor those with the lowest spatial entropy, but ones with intermediate values. Lower H and wcss are desired. However, a tradeoff between these two metrics is usually observed. As already pointed out by Oliver and Webster [24] and was later empirically confirmed by Martin and Boisvert [18], higher cohesion in the multivariate space usually leads to the geographic fragmentation of the clusters. The solution is to combine both metrics and evaluate the results comparatively.

In order to further analyze and validate the geographic connectivity of the clusters, we used the mapping of the spatial continuity of the indicators based on Modena et al. [19]. However, those authors used variograms, which may be noisy in short distances. Here we use correlograms, as these are standardized, which gives more stability to the results, especially regarding short distances. In this method, a binary variable is defined, assuming value 1 for samples within the cluster being evaluated and 0 for all others. The correlogram for this binary variable is then plotted for different lags. Continuous clusters will result in structured correlograms and, fragmented clusters, in noisy correlograms and/or high nugget effects.

Choosing the adequate clustering configuration is a very subjective task. It is important to observe the statistical distributions of the clusters using tools such as histograms, boxplots, and scatter plots. Additionally, comparisons between the defined clustering domains and some geological aspects of the samples (e.g., rock type, alteration patterns) are essential.

## 2.3  *Automatic Classification of New Samples*

In order to perform the automatic classification, first the clustering-labeled dataset must be used to calibrate (test and define the parameters) the classifier. Differently from what happens in unsupervised techniques, validating the performance of a supervised algorithm is straightforward, as the data is already labeled, as it will be demonstrated in the illustration case that follows. In order to do so, the dataset must be divided into the training and test subsets. The quality of the classifier can be assessed with confusion matrices and with the examination of global metrics such as recall, precision, and accuracy. A prevalent and proper validation technique for supervised learning applications is the k-fold cross-validation, which splits the dataset into train and test sets several times, each of these with different sets of data to avoid overfitting.

## 2.4 Workflow

For a better understanding, the proposed workflow can be organized as follows:

(i)   Exploratory data analysis;
(ii)  Data standardization according to

$$Z = \frac{(X - m)}{s} \tag{1}$$

where $Z$ is the standardized value, $X$ is the original value, $m$ is the mean, and $s$, the standard deviation of the original values;

(iii)  Computation of the mean scores of the Silhouette, Calinski-Harabasz, Davies-Bouldin, wcss, and H;
(iv)   Selection of the best scenarios based on the resulting scores from step (iii);
(v)    Verification of the geographic contiguity of the clusters, using the correlograms of the indicators defined by the classes;
(vi)   Visual inspection and statistical analysis of the scenarios selected in step (v);
(vii)  Selection of the best-suited scenario based on the previous steps;
(viii) Calibration of a supervised classifier and automatic classification of new samples.

All methods were run using a Jupyter Notebook, under Windows 10, 64 bit with Python 3.6.5 installed via Anaconda; processor Intel® i7-3.20Ghz, with 24.0 GB RAM.

# 3  Case Study

## 3.1  Exploratory Data Analysis

The techniques mentioned in the previous sections were applied to a three-dimensional isotopic dataset from a phosphate-titanium deposit, containing 19.344 samples assayed for 12 oxides ($P_2O_5$, $Fe_2O_3$, $MgO$, $CaO$, $Al_2O_3$, $SiO_2$, $TiO_2$, $MnO$, $Na_2O$, $K_2O$, $BaO$, and $Nb_2O_5$) and loss on ignition (LoI). The samples also carry

two categorical attributes: weathering patterns (Fig. 2a) and rock types (Fig. 2b).
Figure 3 shows the histograms of the continuous variables, and Fig. 4, the bar charts
with the sample counts of the two categorical variables. Tables 1 and 2 show brief
descriptions of each of the typologies.

The distributions of each continuous variable are quite different, and some of
them show considering asymmetry. Besides, it can be observed that complex relationships are present from the scatterplots of the most relevant variables (Fig. 5).
These complexities and the differences in the scale of the raw variables can lead

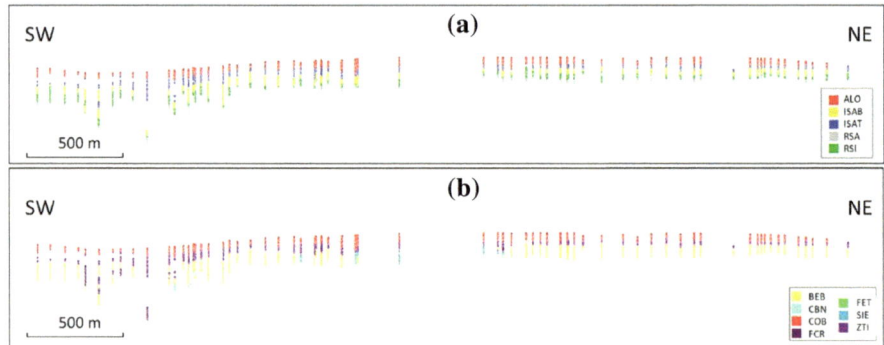

**Fig. 2** Cross-section (N30E) showing samples symbolized by weathering (**a**) and rock type (**b**)

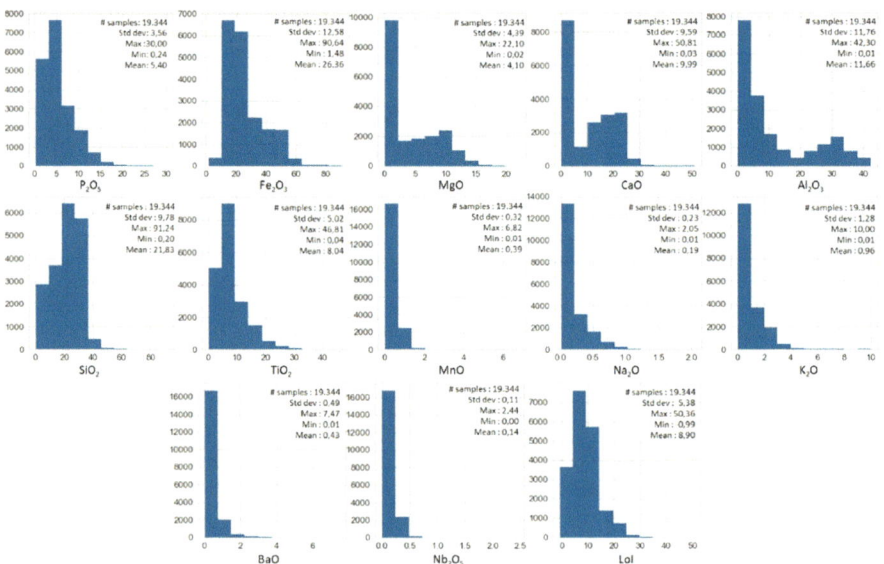

**Fig. 3** Histograms of the continuous variables

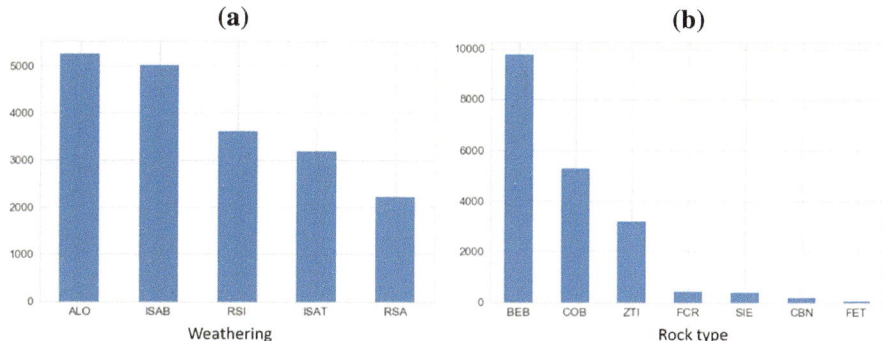

**Fig. 4** Bar charts with the sample counts of the categorical variables

**Table 1** Rock type typologies

| Typology | Description |
| --- | --- |
| COB | Soil cover |
| ZTI | Titanium zone |
| BEB | Bebedourites |
| FCR | Foscorites |
| FET | Foscrete |
| CBN | Carbonatite |
| SIE | Syenites |

**Table 2** Weathering typologies

| Typology | Description |
| --- | --- |
| ALO | Soils, clays and laterites |
| ISAT | Saprolite generally mineralized in titanium |
| ISAB | Saprolite generally mineralized in phosphate |
| RSI | Semi-altered rock |
| RSA | Fresh rock |

to inconsistencies when applying machine learning algorithms, so all continuous variables were standardized according to Eq. 1.

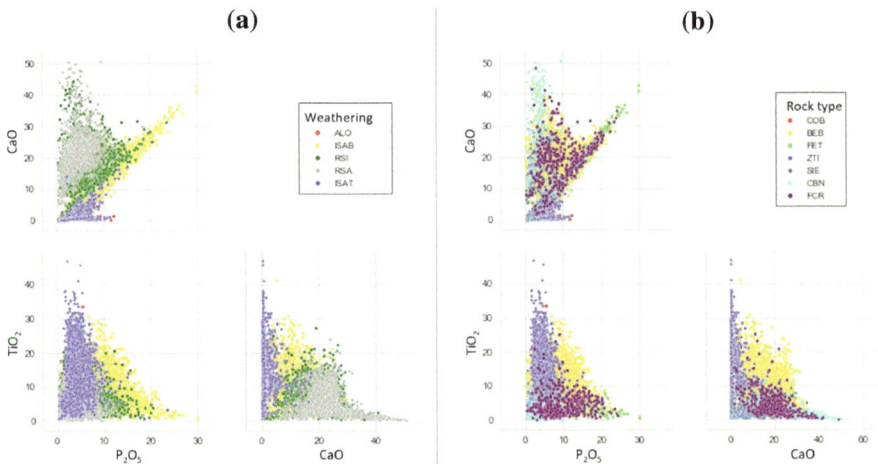

**Fig. 5** Scatter plots of the most relevant continuous variables, with the points colored according to the weathering code and rock type

## 3.2  Applying Cluster Analysis and Verifying the Results

The four clustering algorithms (k-means, hierarchical, *dsclus* and *acclus*) were applied in seven different scenarios, each corresponding to a different number of clusters, from two to seven. The validation methods mentioned in the previous sections were applied in order to evaluate the clustering configurations.

To run the hierarchical and k-means techniques, we used the algorithms available on the Scikit-learn library [26]. For the first, we used the ward's proximity distances and for the latter, the k-means++ option for the centroid initialization, which seeks to maximize the separation between the initial centroids, increasing speed and accuracy.

For *dsclus* and *acclus*, we used the algorithms available on GitHub, hosted in the account mentioned in Martin and Boisvert [18]. The *dsclus* being a clustering ensemble technique, we used 100 realizations, with 20 nearest neighbors and search volume $= (0, 0, 0, 400, 400, 12)$. For *acclus*, 30 nearest neighbors were set, also with a search volume $= (0, 0, 0, 400, 400, 12)$.

Figure 6 shows the results of the multivariate metrics and the spatial entropy for all four algorithms in all scenarios. It becomes evident that the traditional algorithms produce better results than the spatial techniques in the multivariate space, with k-means outperforming the hierarchical method. On the other hand, spatial techniques show better results in the geographic space, with *dsclus* outperforming *acclus* in almost every case. Therefore, it is noticeable that *dsclus* is preferable, as it shows more balanced results between the multivariate and the geographic aspects.

To verify geographic connectivity and multivariate organization simultaneously, the dual space metrics of [18]—wcss and spatial entropy—can be plotted in a scatter plot (Fig. 7). As already stated, low values for both H and wcss, are desirable.

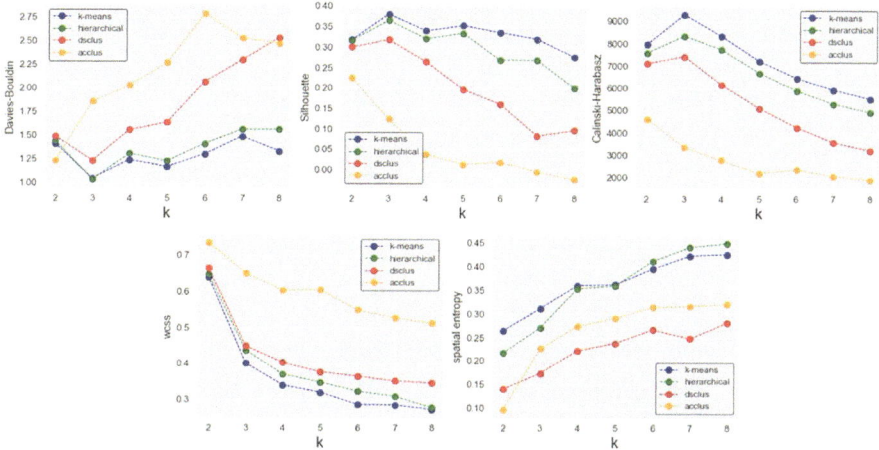

**Fig. 6** Plots showing the variations for Davies-Bouldin, Silhouette, Calinski-Harabasz, wcss and spatial entropy (H) scores, applied to the clustered data as the number of clusters (k) increases

However, it is apparent in the figure that these two metrics are inversely proportional. The solution is to look for intermediate configurations that present a better balance between multivariate and geographic cohesion.

As for choosing the adequate number of clusters (k), intermediate values of all metrics are also preferable. Thus, even though the scenarios with two or three clusters show good scores, these configurations can lead to the statistical mixing of populations, which must be avoided, especially concerning a complex situation with high dimensionality. On the other hand, grouping the data in seven, eight, or even six

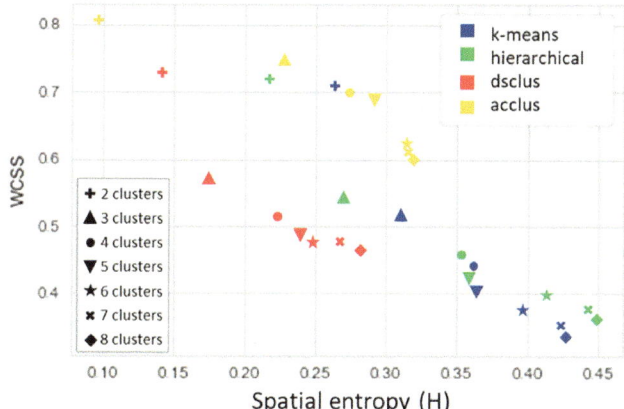

**Fig. 7** Spatial entropy and wcss plotted against each other to simultaneously evaluate the clustering configurations in multivariate and geographic spaces. The color indicates the clustering algorithm, the icon, the number of clusters

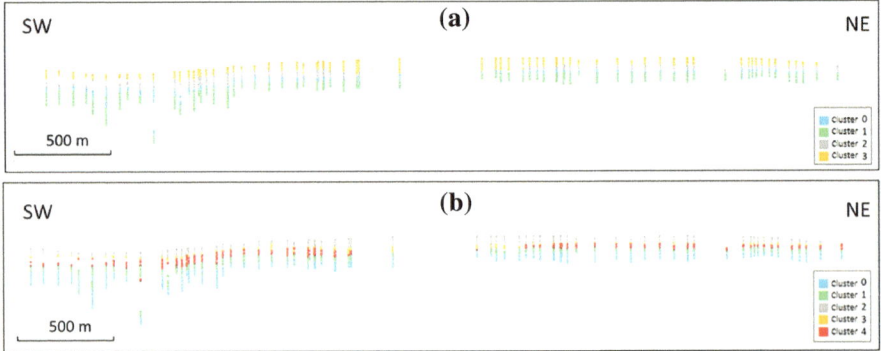

**Fig. 8** Cross-sections are showing part of the dataset, with samples colored by clustering domain (*dsclus*), on the four (**a**) and five-cluster (**b**) scenarios

clusters can take unnecessary complications in the following steps of the modeling process, such as the definition of contours and estimation, as some groups will be somewhat redundant. Therefore, the *dsclus*-clustered data with four or five groups seem to be better options and will be the only scenarios to be considered from now on.

Figure 8 shows the spatial distribution of the samples, colored by clustering code, to understand the clusters' layout better. It becomes clear that the distribution in each case is comparable to the distributions of the weathering patterns and rock type (see Fig. 2), with a dominant horizontal layout.

A statistical evaluation of samples was also performed, and, as can be seen in the boxplots of Figs. 9 and 10, both scenarios show different statistical distributions, depending on the considered variable.

Finally, a sample count was performed to compare clusters and the categorical variables—weathering and rock type—and Fig. 11 presents the bar charts with the results.

## 3.3 Discussions on the Results of the Cluster Analysis

From the statistical analysis and the comparisons between clusters, weathering patterns, and rock types, a description of each cluster can be drawn so that they can compose modeling domains (Tables 3 and 4).

The scenario with four domains can be compared to the model used at the mine site in the case study. What is considered phosphate and titanium ores are saprolite materials comprised in our domains 0 and 2, respectively. The overburden waste, composed of soil, clay, and laterite, corresponds to our domain 3. Altered and fresh

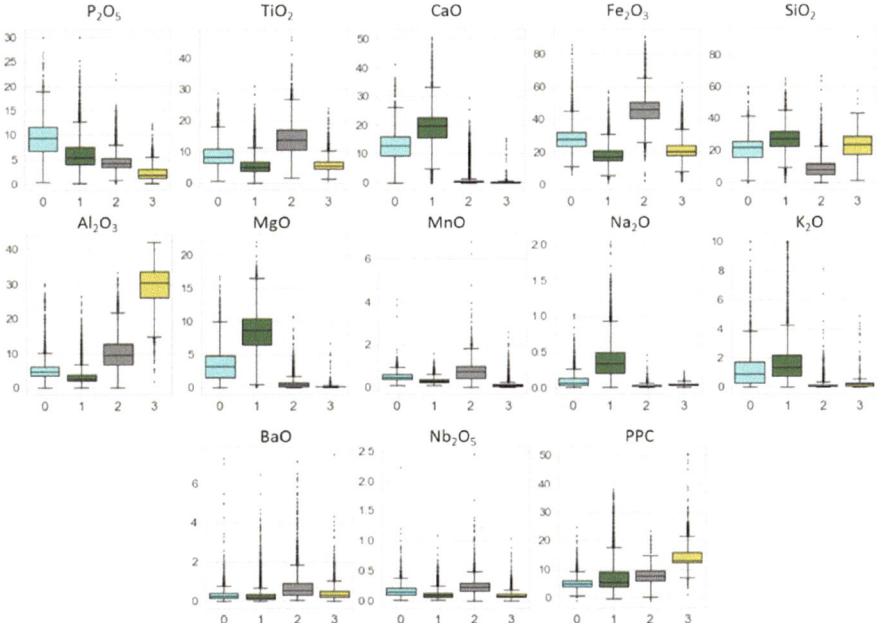

**Fig. 9** Boxplots with the statistical distributions of samples per group in the four-cluster *dsclus* scenario

rocks from the base of the deposit are grouped in our domain 1. In the mine, semi-altered rocks can also be mined as phosphate ore, depending on $P_2O_5$ grades and $CaO/P_2O_5$ ratios.

Although the four-cluster scenario is the most similar to the model practiced in the mine, this resemblance shows some inconsistencies, mainly regarding the rocky materials (RSA and RSI). In the mine, fresh and altered rock materials are included in different modeling domains. This is due to a technical artifact: this way, particularly low grades of $P_2O_5$ and/or high grades of $CaO$ will not contaminate the estimates of blocks classified as an altered rock in the model, which can be mined as ore in certain areas. In the global perspective of the cluster analysis, as conducted in this study, there is no evidence that RSA and RSI should constitute different modeling domains. Besides, the four-cluster configuration implies considerable losses of materials that could be mined as ore (included in our domain 1).

In the five-domain scenario, domain 0 comprises the rocky waste and phosphate-bearing rocks with high carbonate contents. Domain 1 is the phosphate ore, domain 4 is the titanium ore, and domain 2 is the overburden. Domain 3, although mainly composed of overburden, can be an alternate source of titanium. We conclude that this configuration is adequate compared to the four-domain scenario.

Another important observation is that these comparisons between the clustering results and the mine model are subjective evaluations because the geologic classifications (weathering patterns and rock types) are products of the personal interpretations

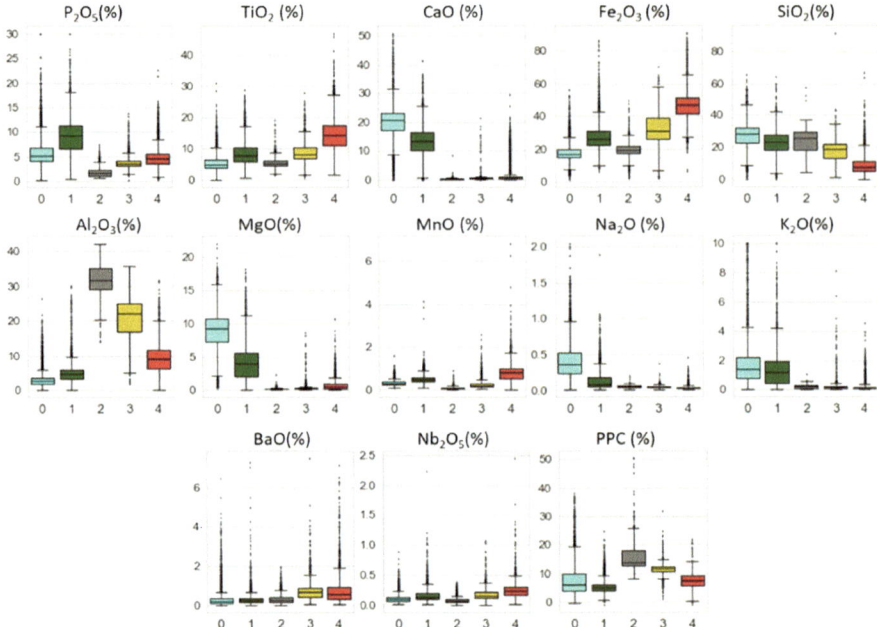

**Fig. 10** Boxplots with the statistical distributions of samples per group in the five-cluster *dsclus* scenario

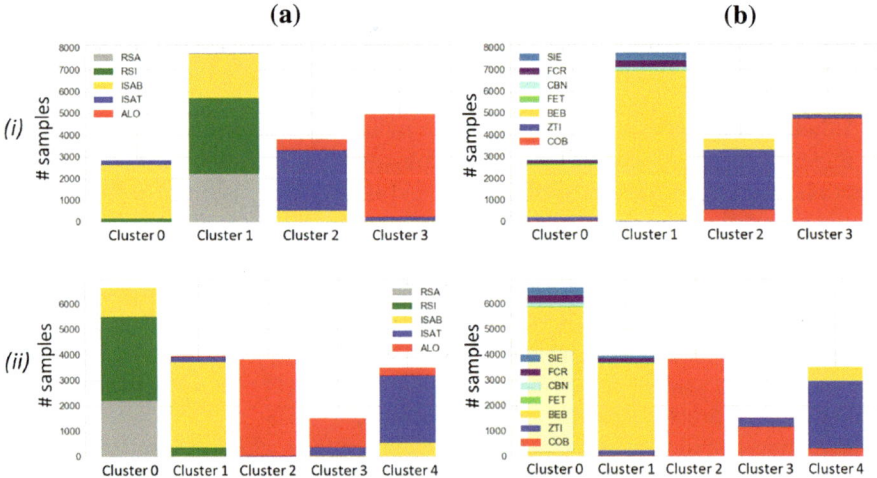

**Fig. 11** Bar charts comparing weathering and rock types to the clusters in the four and five-cluster scenarios obtained with de *dsclus* algorithm

**Table 3** Characterization of each cluster of data in the four-cluster *dsclus* scenario

| Cluster/domain | Description |
|---|---|
| 0 | The main source of phosphate ore. Locally, it can show considerable grades of $TiO_2$, $Fe_2O_3$, CaO and MgO |
| 1 | Rocky materials (altered or fresh) in the base of the deposit. Locally can contain materials with interesting grades of $P_2O_5$ as a possible source of phosphate ore |
| 2 | The main source of titanium ore. May locally contain high grades of $Fe_2O_3$, $Al_2O_3$, MnO and BaO |
| 3 | Overburden (soil, clay, etc.). High grades of $Al_2O_3$ and LoI |

**Table 4** Characterization of each cluster of data in the five-cluster dsclus scenario

| Cluster/domain | Description |
|---|---|
| 0 | Rocky materials (altered or fresh) in the base of the deposit. Locally can contain materials with interesting grades of $P_2O_5$, but also CaO, MgO (carbonates), $Na_2O$, and $K_2O$ (syenite) |
| 1 | The main source of phosphate ore. Locally, can show considerable grades of $TiO_2$, $Fe_2O_3$, $Na_2O$, $K_2O$ (syenites), CaO and MgO (carbonates) |
| 2 | Overburden (soil, clay, etc.). High grades of $Al_2O_3$ and LOI |
| 3 | Mainly composed of overburden materials, but can contain considerable grades of $TiO_2$ |
| 4 | The main source of titanium ore. May locally contain high grades of $P_2O_5$, $Fe_2O_3$, MnO and BaO |

of the company's personnel. Thus, those classifications should not be taken as true data labels, although the matches with the clustering domains show a remarkable similarity.

## 3.4 Supervised Learning Applied to the Automatic Classification of New Samples

Once the most suitable scenario is chosen, the clustered data can be used to train a supervised classifier to be applied in the classification of new samples. Many algorithms can be used, such as k-nearest neighbors, decision trees, and random forests. In this paper, the latter was chosen.

First, the 19.344 labeled dataset was segmented into stratified training and testing sets in a proportion of 85%–15% (17.252 and 2.902 samples, respectively). As this is a case where the spatial configuration is important in defining clusters, the xyz coordinates were also used as input variables, along with all continuous attributes.

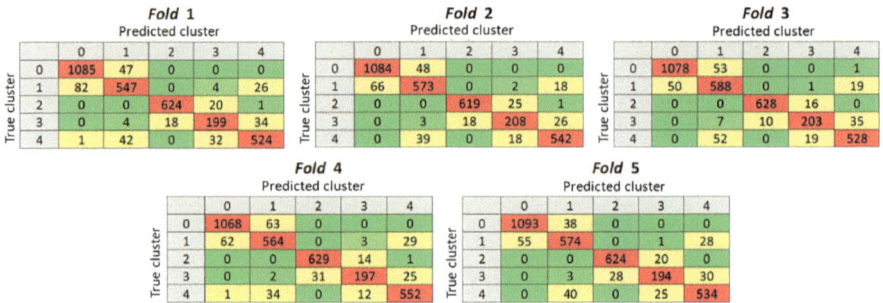

**Fig. 12** Confusion matrices from each of the five folds from the stratified fivefold cross validation applied to evaluate the Random Forest classifier

The input variables were standardized according to Eq. 1. The 'Randomized Search CV' (Pedregosa et al., 2012) method was used to define the best parameters of the Random Forest classifier. Fivefold stratified cross-validation was applied, and results are shown in Figs. 12 and 13.

The model was run with the 2.902-sample test set. It was verified that the results (Fig. 14) are consistent to those observed in the fivefold stratified cross validation, attesting that there is no under or overfitting in the model. The statistical affinities between some groups lead to some misclassifications, but results are acceptable, with a 92% overall accuracy.

As a final evaluation, the statistical distribution and the visual inspection of the automatically classified samples from the test set were checked. Figures 15 and 16 show that the results are consistent (see Figs. 8b and 10 for comparison).

# 4 Conclusions

Although very efficient in many cases, the application of traditional clustering algorithms is quite limited in modeling mineral resources since they only consider data relationships in the multivariate space, neglecting their geographic distribution. Thus, techniques that also consider the geographic distribution of samples are more appropriate, as demonstrated in the case study.

Parameterization and the validation of the clustering results are still complex decisions and entirely subjective. Therefore, despite being very effective in the decision-making process, those methods are not yet fully automated, requiring specialized knowledge and good judgment.

The geological characteristics of a mineral deposit should guide the classification of the samples. However, traditionally, these characteristics lead to subjective classifications, resulting from personal interpretation in the data acquisition phase. Such classifications should therefore not be given as unquestionable labels, and differences

**Fig. 13** Global metrics for each of the five folds from the stratified fivefold cross-validation

| | Cluster | Precision | Recall | F1-score | Support |
|---|---|---|---|---|---|
| Fold 1 | Cluster 0 | 0.93 | 0.96 | 0.94 | 1132 |
| | Cluster 1 | 0.85 | 0.83 | 0.84 | 659 |
| | Cluster 2 | 0.97 | 0.97 | 0.97 | 645 |
| | Cluster 3 | 0.78 | 0.78 | 0.78 | 255 |
| | Cluster 4 | 0.90 | 0.87 | 0.88 | 599 |
| Fold 2 | Cluster 0 | 0.94 | 0.96 | 0.95 | 1132 |
| | Cluster 1 | 0.86 | 0.87 | 0.86 | 659 |
| | Cluster 2 | 0.97 | 0.96 | 0.96 | 645 |
| | Cluster 3 | 0.82 | 0.82 | 0.82 | 255 |
| | Cluster 4 | 0.92 | 0.90 | 0.91 | 599 |
| Fold 3 | Cluster 0 | 0.96 | 0.95 | 0.95 | 1132 |
| | Cluster 1 | 0.84 | 0.89 | 0.86 | 659 |
| | Cluster 2 | 0.98 | 0.98 | 0.98 | 645 |
| | Cluster 3 | 0.85 | 0.80 | 0.82 | 255 |
| | Cluster 4 | 0.91 | 0.88 | 0.89 | 599 |
| Fold 4 | Cluster 0 | 0.94 | 0.94 | 0.94 | 1132 |
| | Cluster 1 | 0.85 | 0.86 | 0.85 | 659 |
| | Cluster 2 | 0.95 | 0.98 | 0.96 | 645 |
| | Cluster 3 | 0.87 | 0.77 | 0.82 | 255 |
| | Cluster 4 | 0.91 | 0.92 | 0.91 | 599 |
| Fold 5 | Cluster 0 | 0.95 | 0.97 | 0.96 | 1132 |
| | Cluster 1 | 0.88 | 0.87 | 0.87 | 659 |
| | Cluster 2 | 0.96 | 0.97 | 0.96 | 645 |
| | Cluster 3 | 0.81 | 0.76 | 0.78 | 255 |
| | Cluster 4 | 0.90 | 0.89 | 0.89 | 599 |

| Domain | Precision | Recall | F1-score | Support |
|---|---|---|---|---|
| 0 | 0.95 | 0.96 | 0.95 | 956 |
| 1 | 0.89 | 0.89 | 0.89 | 637 |
| 2 | 0.97 | 0.97 | 0.97 | 593 |
| 3 | 0.85 | 0.82 | 0.83 | 232 |
| 4 | 0.91 | 0.91 | 0.91 | 484 |
| | | Accuracy 0.93 | | |

| | | Predicted Cluster | | | | |
|---|---|---|---|---|---|---|
| | | 0 | 1 | 2 | 3 | 4 |
| True cluster | 0 | 918 | 38 | 0 | 0 | 0 |
| | 1 | 49 | 565 | 0 | 1 | 22 |
| | 2 | 0 | 0 | 574 | 18 | 1 |
| | 3 | 0 | 4 | 17 | 190 | 21 |
| | 4 | 0 | 29 | 0 | 15 | 440 |

**Fig. 14** Global metrics and confusion matrix for the validation of the Random Forest classifier when applied to the 2.902-sample test set

should arise compared to the results of clustering algorithms. It is evident that the geological characteristics must be reflected in the clusters in some way.

Once the best scenario has been defined, the codes of each group can be fed as labels in supervised learning algorithms (e.g., decision trees, random forests, k-nearest neighbors) for the calibration of mathematical models for the automatic

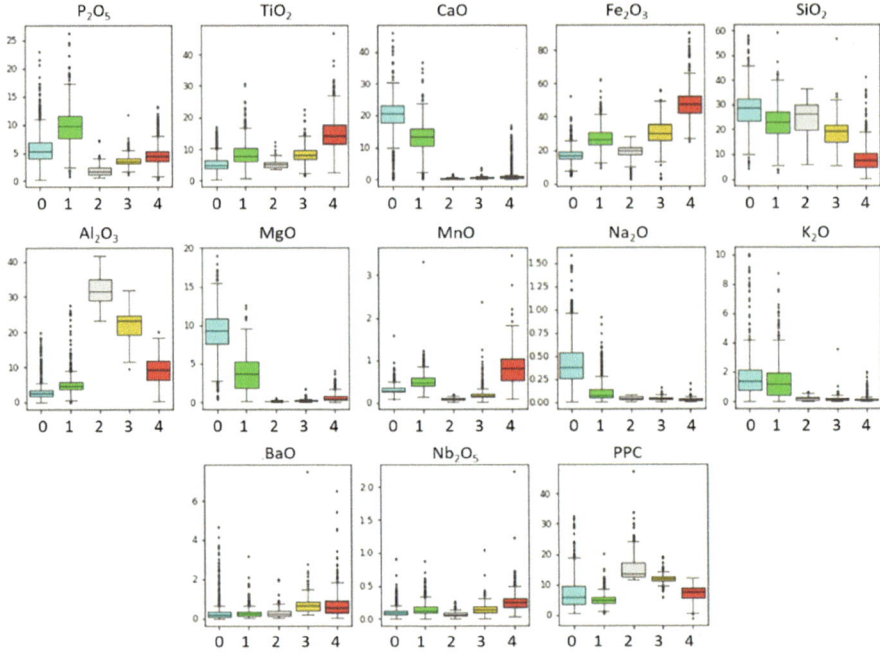

**Fig. 15** Boxplots show the statistical distributions of the 2.902 samples labeled with the Random Forest classifier

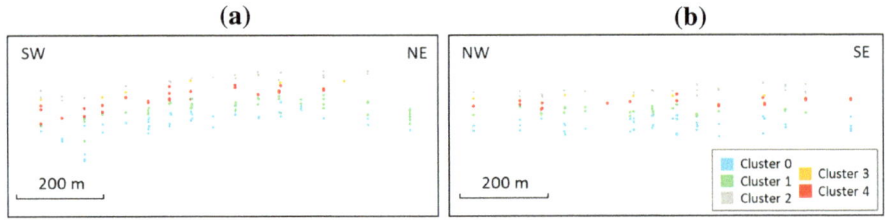

**Fig. 16** Cross sections show the spatial distributions of some of the 2.902 samples labeled with the Random Forest classifier

classification of new samples. Periodically, the complete analysis should be revisited with all samples, and the supervised classifier, updated.

One of the most significant advantages of applying machine learning to mining problems is their ability to make measurements in a multidimensional space, using complex mathematical relationships, which is practically impossible for a human analyst. Apart from the subjectivity of the validation and final decision, this automation significantly increases reproducibility in the modeling processes, which is essential in evaluating mineral resources, especially for audition purposes.

**Acknowledgements** The authors would like to thank the Mineral Exploration and Mining Planning Laboratory (LPM) at the Federal University of Rio Grande do Sul (UFRGS), for providing the necessary conditions for developing this work. Luiz Englert Foundation (FLE), the Coordination for the Improvement of Higher Education Personnel (Capes) and the National Council for Scientific and Technological Development (CNPq) are acknowledged for their financial support. We would also like to thank Dr. Ryan Martin for giving access to his codes and Mosaic Fertilizantes for providing the data.

# References

1. Ambroise, C., Dang, M., Govaert, G.: Clustering of spatial data by the EM algorithm. In: Proceedings of GEOEV I—Geostatistics for Environmental Applications, pp. 493–504 (1997). https://doi.org/10.1007/978-94-017-1675-8_40
2. Calinski, T., Harabasz, J.: A dendrite method for cluster analysis. Commun. Stat. Theory Methods **3**(1), 1–27 (1974). https://doi.org/10.1080/03610927408827101
3. D'Urso, P., Vitale, V.: A robust hierarchical clustering for georeferenced data. Spatial Stat., **35** (2020)
4. Davies, D.L., Bouldin, D.W.: A cluster separation measure. IEEE Trans. Pattern Anal. Mach. Intell. **2**, 224–227 (1979). https://doi.org/10.1109/TPAMI.1979.4766909
5. Dempster, A.P., Laird, N.M., Rubin, D.B.: Maximum likelihood from incomplete data via the EM algorithm. J. Roy. Stat. Soc.: Ser. B (Methodol.) **39**(1), 1–22 (1977). https://doi.org/10.1111/j.2517-6161.1977.tb01600.x
6. Drumond, D. A.: Estimativa e classificação de variáveis geometalúrgicas a partir de técnicas de aprendizado de máquinas. Universidade Federal do Rio Grande do Sul (2019). Retrieved from https://www.lume.ufrgs.br/handle/10183/202480
7. Ester, M., Kriegel, H.-P., Sander, J., Xu, X.: A density-based algorithm for discovering clusters in large spatial databases with noise. In: Proceedings of the 2nd International Conference on Knowledge Discovery and Data Mining, pp. 226–231 (1996). https://www.aaai.org/Papers/KDD/1996/KDD96-037.pdf
8. Fouedjio, F.: A hierarchical clustering method for multivariate geostatistical data. Spatial Stat. **18**, 333–351 (2016). https://doi.org/10.1016/j.spasta.2016.07.003
9. Fouedjio, F.: A spectral clustering approach for multivariate geostatistical data. Int. J. Data Sci. Anal. **4**(4), 301–312 (2017). https://doi.org/10.1007/s41060-017-0069-7
10. Fouedjio, F.: A spectral clustering method for large-scale geostatistical datasets. In: Proceedings of the Thirteenth International Conference in Machine Learning and Data Mining in Pattern Recognition, vol. 10358 LNAI, pp. 248–261 (2017). https://doi.org/10.1007/978-3-319-62416-7_18
11. Fouedjio, F.: Clustering of multivariate geostatistical data. Wiley Interdiscip. Rev. Comput. Stat. 1–13 (2020). https://doi.org/10.1002/wics.1510
12. Fouedjio, F., Hill, E.J., Laukamp, C.: Geostatistical clustering as an aid for ore body domaining: case study at the Rocklea Dome channel iron ore deposit, Western Australia. Appl. Earth Sci. Trans. Inst. Mining Metallurgy **127**(1), 15–29 (2018). https://doi.org/10.1080/03717453.2017.1415114
13. Getis, A., Ord, J.K.: The analysis of spatial association by use of distance statistics. Geogr. Anal. **24**(3), 189–206 (1992). https://doi.org/10.1111/j.1538-4632.1992.tb00261.x
14. Gülcan, E., Gülsoy, Ö.Y.: Performance evaluation of optical sorting in mineral processing—A case study with quartz, magnesite, hematite, lignite, copper and gold ores. Int. J. Miner. Process. (2017). https://doi.org/10.1016/j.minpro.2017.11.007
15. Journel, A.G., Huijbregts, C.: Mining geostatistics. Academic Press Limited, London (1978)

16. Lishchuk, V., Lund, C., Ghorbani, Y.: Evaluation and comparison of different machine-learning methods to integrate sparse process data into a spatial model in geometallurgy. Miner. Eng. **134**, 156–165 (2019). https://doi.org/10.1016/j.mineng.2019.01.032

17. MacQueen, J.: Some methods for classification and analysis of multivariate observations. In: Proceedings of the Fifth Berkeley Symposium on Mathematical Statistics and Probability, vol. 1, pp. 281–296 (1967).

18. Martin, R., Boisvert, J.: Towards justifying unsupervised stationary decisions for geostatistical modeling: Ensemble spatial and multivariate clustering with geomodeling specific clustering metrics. Comput. Geosci. **120**, 82–96 (2018). https://doi.org/10.1016/j.cageo.2018.08.005

19. Modena, R.C.C., Moreira, G. de C., Marques, D.M., Costa, J.F.C.L.: Avaliação de técnicas de agrupamento para definição de domínios estacionários com o auxílio de geoestatística. In: Proceedings of the Twentieth Mining Symposium at the Fifth ABM Week, pp. 91–100 (2019). São Paulo. https://doi.org/10.5151/2594-357x-33405

20. Moreira, G.C.: Análise de agrupamento aplicada à definição de domínios de estimativa para a modelagem de recursos minerais. Universidade Federal do Rio Grande do Sul (2020). Retrieved from https://lume.ufrgs.br/handle/10183/212457

21. Moreira, G.C., Costa, J.F.C.L., Marques, D.M.: Defining geologic domains using cluster analysis and indicator correlograms: a phosphate-titanium case study. Appl. Earth Sci. **129**(4), 176–190 (2020). https://doi.org/10.1080/25726838.2020.1814483

22. Nakhaei, F., Mosavi, M.R., Sam, A., Vaghei, Y.: Recovery and grade accurate prediction of pilot plant flotation column concentrate: Neural network and statistical techniques. Int. J. Miner. Process. **110**, 140–154 (2012). https://doi.org/10.1016/j.minpro.2012.03.003

23. Niquini, F.G.F., Costa, J.F.C.L.: Mass and metallurgical balance forecast for a zinc processing plant using artificial neural networks. Nat. Resour. Res. (2020). https://doi.org/10.1007/s11053-020-09678-4

24. Oliver, M.A., Webster, R.: A geostatistical basis for spatial weighting in multivariate classification. Math. Geol. **21**(1), 15–35 (1989). https://doi.org/10.1007/BF00897238

25. Ord, J.K., Getis, A.: Local spatial autocorrelation statistics: distributional issues and an application. Geogr. Anal. **27**, 286–306 (1995). https://doi.org/10.1111/j.1538-4632.1995.tb00912.x

26. Pedregosa, F., Varoquaux, G., Gramfort, A., Michel, V., Thirion, B., Grisel, O., et al.: Scikit-learn: machine learning in python. J. Mach. Learn. Res. **12**, 2825–2830 (2011)

27. Romary, T., Ors, F., Rivoirard, J., Deraisme, J.: Unsupervised classification of multivariate geostatistical data: two algorithms. Comput. Geosci. **85**, 96–103 (2015). https://doi.org/10.1016/j.cageo.2015.05.019

28. Romary, T., Rivoirard, J., Deraisme, J., Quinones, C., Freulon, X.: Domaining by clustering multivariate geostatistical data. In: Geostatistics Oslo, pp. 455–466 (2012). https://doi.org/10.1007/978-94-007-4153-9_37

29. Rossi, M.E., Deutsch, C.V.: Mineral resource estimation. Mineral Resource Estimation. Springer Science & Business Media (2014). https://doi.org/10.1007/978-1-4020-5717-5

30. Rousseeuw, P.J.: Silhouettes: A graphical aid to the interpretation and validation of cluster analysis. J. Comput. Appl. Math. **20**(C), 53–65 (1987). https://doi.org/10.1016/0377-0427(87)90125-7

31. Samuel, A.L.: Some studies in machine learning using the game of checkers. IBM J. Res. Dev. **3**(3), 210–229 (1959)

32. Scrucca, L.: Clustering multivariate spatial data based on local measures of spatial autocorrelation. Quaderni del Dipartimento di Economia, Finanza e Statistica **20**(1), 1–25 (2005). http://www.ec.unipg.it/DEFS/uploads/spatcluster.pdf

33. Shu, L., Osinski, G.R., McIsaac, K., Wang, D.: An automatic methodology for analyzing sorting level of rock particles. Comput. Geosci. **120**, 97–104 (2018). https://doi.org/10.1016/j.cageo.2018.08.001

34. Sokal, R.R., Sneath, P.H.A.: Principles of numerical taxonomy. J. Mammal. **46**(1), 111 (1965). https://doi.org/10.2307/1377831

# Addressing Application Challenges with Large-Scale Geological Boundary Modelling

**Adrian Ball, John Zigman, Arman Melkumyan, Anna Chlingaryan, Katherine Silversides, and Raymond Leung**

**Abstract** For banded iron formation-hosted deposits accurate boundary modelling is critical to ore-grade estimation. Key to estimation fidelity is the accurate separation of the different domains within the ore body, requiring modelling of the boundaries between domains. This yields both theoretical and application challenges. We present a series of solutions for application challenges that arise when modelling large-scale boundaries employing a composition of Gaussian Process models on exploration and production hole data. We demonstrate these in the banded iron formation-hosted iron ore deposits in the Hamersley Province of Western Australia. We present solutions to several challenges: the inclusion of information derived from a geologist-defined boundary estimate to incorporate domain knowledge in data sparse regions, the incorporation of unassayed production holes that are implicitly defined as waste to augment production hole assay data, and a more holistic method of defining regional bounds and spatial rotations for Gaussian Process modelling of local spaces. Solution are evaluated against a range of metrics to show performance improvements over the

A. Ball (✉) · A. Melkumyan · A. Chlingaryan · R. Leung
Australian Centre for Field Robotics, The University of Sydney, Rose Street Building (J04), University of Sydney, NSW 2006, Australia
e-mail: adrian.ball@sydney.edu.au

A. Melkumyan
e-mail: arman.melkumyan@sydney.edu.au

A. Chlingaryan
e-mail: anna.chlingaryan@sydney.edu.au

R. Leung
e-mail: raymond.leung@sydney.edu.au

J. Zigman
Toronto, Canada

K. Silversides
Australian Centre for Field Robotic, The University of Sydney, Camperdown, Australia
e-mail: katherine.silversides@sydney.edu.au

© The Author(s) 2023
S. A. Avalos Sotomayor et al. (eds.), *Geostatistics Toronto 2021*, Springer Proceedings in Earth and Environmental Sciences, https://doi.org/10.1007/978-3-031-19845-8_17

221

manually performed estimation by an expert geologist of the boundaries delineating the ore body domains. Reconcilliation scores are used for evaluating the quality of predicted domain boundaries against measured production data. The predicted and in situ surfaces are also qualitatively evaluated against production data to ensure that the models were evaluated to be geologically sound by an expert in the field. In particular, better fidelity is shown when separating mineralised and non-mineralised ore, consequently improving the estimation of the ore-grades present in the mine site.

**Keywords**  Geologic domains · Sparse data · Resource estimation

# 1  Introduction

When mining, accurate ore grade estimation is critical as it influences mine planning [1], logistics, and product reliability. In stratified ore deposits—such as banded iron formation (BIF) hosted iron ore deposits—accurate boundary estimation is a prerequisite for high fidelity, accurate ore grade estimations. Poor boundary modelling can result in the inclusion of ore into a waste region or vice versa. This has a deleterious effect on ore grade estimates, due to either lower ore recovery or ore dilution and is the equivalent of the inclusion of bad/incorrect data into the model. Reduced fidelity can result from either poor boundary position estimates or by too coarse a tessellated model surface. The coarseness of the tessellation introduces a trade-off between computational costs and model fidelity.

While much work has been done on implicit modelling methods for geological boundaries [2, 3], considerably less work has been done on probabilistic methods for modelling boundaries. Neves et al. [4] also considered using geochemical data rapidly obtained from portable XRF devices to update potentially out-of-date grade estimates in what is termed as real-time mining. Their proposed method models the uncertainty of XRF measurements by considering their conditional distribution using confident laboratory assays (hard and sparse data) that derive from exploration holes. A distinguishing feature of our paper is that we focus on the location of geological boundaries rather than grade estimation per se; additionally our approach is based on Gaussian Processes rather than stochastic simulation. This study builds on previous work related to the creation of probabilistic boundaries generated from multiple data types (each having a different levels of noise). The creation of easily updatable boundaries as more data becomes available, builds on the foundation of the Gaussian Processes (GP) probabilistic boundary estimation framework described in [5, 6] and proposes several changes to address issues that arise from a tile-based implementation. The main issues to be resolved are the tendency of GP overfitting and tiling artefacts. The former produces contorted surfaces (unreasonable boundary estimates) particularly in region devoid of input or labelled data. This paper demonstrates that supplying a priori data that properly constrains the solution space (e.g. conservatively indicating where a boundary should not occur) can alleviate boundary distortion. The latter is a manifestation of boundary effects, which occurs when local inference regions have different rotations due to variations in the overall directional trend of the surface and there is no smoothness guarantee for adjacent tile-regions. There are many possible solutions to this problem, one is to introduce a form of weighted transition between adjacent regions. While we focus on one boundary estimation process in this work, some processes (defined below) can be applied to other boundary estimation and update models [7, 8]. These contributions are demonstrated on the banded iron formation-hosted iron ore deposits in the Hamersley Province of Western Australia.

The specific contributions of this work are as follows: (1) The inclusion of a priori data, allowing for the incorporation of domain expertise into the boundary modelling process, preventing the generation of some surface artefacts. (2) A heuristic for the labelling of unassayed production holes, improving boundary modelling accuracy. This further increases the incorporation of domain expertise into the modelling process through the augmentation of available data. (3) The conversion of a set of local rotation calculations defined in [5] to a global rotation model. This allows for the interpolation and extrapolation of rotation transformations across locally modelled sub-regions, providing correlation between sub-region rotations, and preventing artefacts from being introduced by the modelling process.

## 2 Geology

The data used in this study is from two typical Brockman style BIF hosted iron ore deposits from the Hamersley Region in Western Australia. The Brockman Iron Formation contains two sequences of interbedded BIF and shale bands, the Joffre and Dales Gorge Members, as well as two sequences dominated by shale, chert, and/or carbonate bands, the Yandicoogina and Whaleback Shale Members [9, 10]. In some localised areas the BIF in the Joffre and/or Dales Gorge Members has been enriched to form a high grade iron ore [9, 11, 12]. These deposits contain two distinct types of boundaries, stratigraphic and mineralisation. The stratigraphic boundaries are those that follow the bedding of the sequences, either between two members or internal boundaries between sub-units within a member. These are often designated as occuring at specific shale bands. These boundaries define regions with different source rock, which controls the type of ore produced and therefore some of its physical properties. The other type of boundary is related to the mineralisation. These boundaries indicate the areas that have been impacted by geological events after the source rocks were deposited. Examples include where sections of the BIF have been enriched to form iron ore, or where ore quality has been reduced by a hydration overprint. In this study, the stratigraphic boundary is an internal boundary within the Dales Gorge Member, and the mineralisation boundary is located at the base of the ore where it transitions into unenriched BIF.

The data available consisted of exploration drill holes and production blast holes. The exploration holes are spaced ~50 m apart, and are labelled in 2 m intervals.

These labels were added manually by geologists based on the chemical assays and geophysical logging. The labels provided information on both the stratigraphy and the mineralisation, and therefore these holes were used for both boundaries. The blast holes are much more closely spaced, 5–10 m apart, and 10–12 m deep. Each blast hole was given a grade label based on a single chemical assay. This only provided information for the mineralisation boundary, not the stratigraphic boundary. As there was only a single label, it was assigned to the midpoint of the hole.

# 3   Gaussian Processes

In this work, Gaussian Processes are used as a probabilistic non-parametric regression technique. Formally, GPs are a collection of random variables, any finite number of which have a joint Gaussian distribution [13]. A GP is completely defined by its mean function, $m(\mathbf{x})$, and covariance function, $k(\mathbf{x}, \mathbf{x}')$, of a real process $f(\mathbf{x})$ as

$$m(\mathbf{x}) = \mathbb{E}[f(\mathbf{x})],$$
$$k(\mathbf{x}, \mathbf{x}') = \mathbb{E}[(f(\mathbf{x}) - m(\mathbf{x}))(f(\mathbf{x}') - m(\mathbf{x}'))^T],$$

allowing for the GP to be written as

$$f(\mathbf{x}) \sim \mathcal{GP}(m(\mathbf{x}), k(\mathbf{x}, \mathbf{x}')).$$

In this implementation, GPs are used to compute the mean and variance for each point within a regular 3D mesh of points that cover the region of interest. Details relating to the implementation of GPs to this work will be covered in the relevant sections. We encourage the interested reader to refer to [6, 13] for a deeper explanation of GPs.

# 4   A Priori Data

Automated boundary estimation models are generally data driven [14] and contain minimal information that exploits geological domain expertise. Examples of domain expertise include the understanding of the relationship between different surfaces—especially stratigraphic surfaces—and the trend of surfaces outside of the data range [15, 16]. By including a priori data, that is, *data that encapsulates geological expertise*, it is possible to incorporate domain knowledge about the underlying surface being modelled, and its relationship to other surfaces. This knowledge is in the form of a constraint defining where regions are above the surface and below the surface, rather than where the surface is. Perez et al. [17] define high-order training images and present a way of evaluating those against the data seen, here we derive constraint data positioned above or below the data that defines the transition from below to above the surface. The inclusion of this data informs the boundary modelling process and assists in the generation of a boundary estimate that has an appropriate trend in the absence of data.

Further to this, the presence of a priori data will also prevent surface artefacts that can result from GP based boundary estimates [5]. These surface artefacts arise when inferring model values at data sparse locations, due to the tendency of the GP mean function to trend towards zero in these spaces. A manifestation of surface artefacts and the management of it is shown in Fig. 1. In Fig. 1a, the boundary has been estimated without the inclusion of any a priori data. Marching cubes is used to approximate a surface separating the estimates above 0.5 (considered above the surface) from those below 0.5 (below the surface). Using a GP estimation points that are significantly far from data will drift back to the mean, in this case 0. This can result in the introduction of a fictitious isosurface sufficiently far from data where the estimation drifts back below the 0.5 contour level. This tendency causes artificial structures to be introduced in the absence of data, which is a serious problem. The appearance of false surfaces can lead to erroneous interpretations as it incorrectly indicates the location of the boundary. Inclusion of the a priori data prevents the GP mean from tending towards zero in the modelling space at some distance from the provided data. This prevents the generation of surface artefacts, as shown in Fig. 1b.

These a priori points are not included when training a model, only when inferencing from the model across the region of interest. The a priori points represent geological expertise, not actual data, and so are useful in indicating where a surface is not (which complements the spatial region of where the surface may be). By only including the a priori data in the inferencing step, deference is given to physical observations of the region through exploration hole (or other) data. When modelling, the a priori data is placed 'sufficiently far' from the actual data, so as to only guide the generation of a boundary estimate, rather than specify it. The a priori data is dithered to reduce ripple like effects and the associated noise is increased from that of the measured observations.

The a priori data can be defined either through a computational policy mechanism or by utilizing a geological estimate of the boundary. In the case of a computational policy used in the results shown in this paper the introduced data covers gaps above or below real labelled data amount a surface can diverge is constrained. Where a pre-existing surface is used points substantially above or below the surface are used so influence mesh estimations where there is no close by actual data. In either case this data serves as a guide between which a surface is approximated. How close the a priori data need be is a function of the length scales learnt in each sub-region modelled.

The examples shown in this paper contain auto-generated data, where the distance from the actual data is based on the length scales after the section rotation has been applied. However, the distance between the a priori and actual data, and the density, shape, and regularity of the a priori point cloud are all parameters tunable by a geologist, based on the type of information that they wish to encode into the

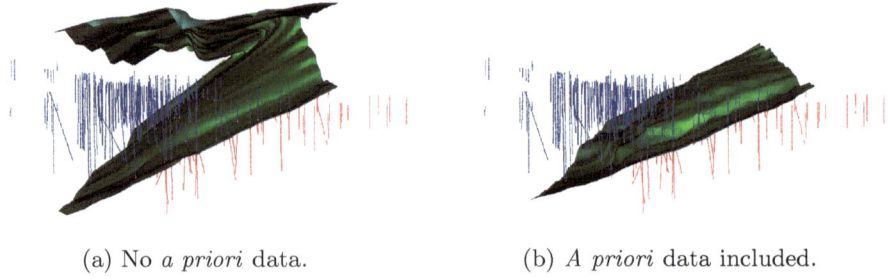

(a) No *a priori* data.                              (b) *A priori* data included.

**Fig. 1** A comparison of two boundary estimates. **a** without using a priori data. **b** when a priori data is included (see Fig. 2). Blue points indicated data labelled as being above the boundary, while red points are labelled as below the boundary

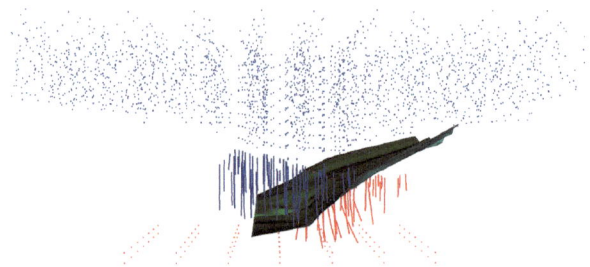

**Fig. 2** Example a priori point clouds can be seen above (blue) and below (red) the available exploration hole data. The a priori is extended to the upper limit of the modelled section and beyond the easting and northing limits of the region to reduce artifacts near the edges

modelling process. An example set of a priori data is shown in Fig. 2. This is a subset of the a priori data that was used to generate the improved surface shown in Fig. 1b.

# 5   Model Building

## 5.1   *Spatial Rotations*

In the boundary modelling method proposed by Ball et al. [5], a global region (with coordinates in ENU) is divided into an overlapping set of local sub-regions. Each local sub-region has an approximate trend direction for the boundary computed for that region. That local mine-space region is rotated into an estimation space so that the nominal trend direction of the surface within that estimation space is horizontal. Each estimation space is a separate local GP model. A rotation matrix for each sub-region was obtained via principal component analysis (PCA). These matrices were calculated from the ore-to-waste transition points provided in exploration hole data within some defined neighbourhood of each local sub-region.

**Fig. 3** Proposed change to the boundary modelling process presented in [5]. Steps listed outside of the blue box are performed once for the entire modelling space. The steps listed inside the blue box are performed once per local sub-region. We propose to move the calculation of local rotation transformations from the local sub-region level to the global model level

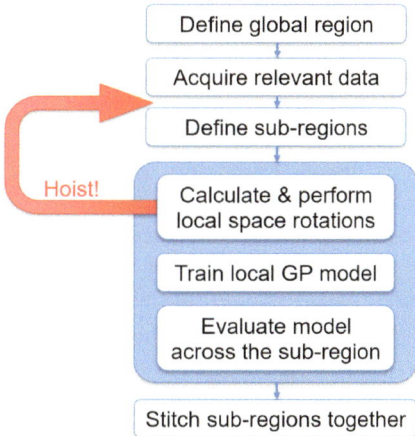

Learning local rotations in the above manner can lead to significant changes between regions due to the inclusion or exclusion of a small number of data points. It is not possible to use PCA for localized rotations using the entire set of data. In comparison, a GP can model rotations where the dependence on the data is a function of the distance from the point of interest and the learnt parameters, i.e. all data is considered by the local data has a greater impact. The consequence is that the inclusion or exclusion of a transition point from determining PCA rotation introduces a stepwise change in the computed rotation. That stepwise change can be significant. To address this the method presented in [5] using PCA is replaced with a GP model for rotations the deflection of the normal from vertical is modelled using the computed normals for the surface at the transition points. This allows a continuously varied estimate of the normal in the mining space being modelled, see Fig. 3.

While PCA rotations can provide significant rotation changes based on inclusion and exclusion of data, it is also true that there may be significant rotation direction differences produced by a GP model of the rotations. Those effects are due to extrapolation of the data (rather than interpolation in the inner regions), the overlap of the regions and how regions and the function used to produce the values for each of the mesh points from the overlapping regions, discussed in Sect. 5.2.

### 5.1.1 Rotational Model Construction

Observation points ENU (Easting, Northing, Up) for the model occurred at the boundary transition point down each exploration hole. At each down-hole transition point, $t_i$, the closest $n$ transition points were used to calculate a rotation matrix, $R_i$, using PCA. The difference in the transition point, $t_i$, of mining sub-region, and the point resulting from the inverse transform of the corresponding point in the estimation space plus an upward unit vector approximates the normal in the mining sub-region, yield an approximate normal in the mining space:

$$t_i - R_i^{-1}(R_i t_i + (0, 0, 1))$$

Note that as $n$ increases, the smoothness of this rotation space also increases. The $x$-, $y$-, and $z$-component vectors of the surface normal unit vector were then used as observation values for three different GP models.

A GP model was fitted for each Easting, Northing and Up of the unit normals at each transition point in the mining space (ENU). This allowed the estimation of a normal at an point within the mining space. The estimates for Easting, Northing and Up parts of the normal are normalized to ensure a unit normal. Given a normal vector of $(e_j, n_j, u_j)$ a rotation matrix is computed such that $(e_j, n_j, u_j) = R_j(0, 0, 1)$, i.e. the vertical unit normal in the modelling space when rotated into back into the mining space matches the mining space normal. Modelling unit normal rotations using polar coordinates were discounted as those introduce multiple values and, consequently, training ambiguities at angles near multiples of $2\pi$.

## 5.2 Region Overlap

Each sub-region is nominally of fixed size and mesh resolution. A region is computed with an overlapping set of transition and a priori data and 3D mesh of estimation points (in ENU mining space). However, the rotations applied in adjacent regions may differ. Furthermore, the length scales learnt for each regions model based on transition data may vary. The 3D mesh of estimation points computed that overlap between regions are merged into single values at each of those points. There are several approaches to computing the merged value of different regions. The methods explored include various methods of weighted averaging where the weights are based on:

- Scaled distance from start to edge of the overlap, i.e. the first point over overlap starting from the centre of a region is given a weighting of close to 1, the furthest point of overlap starting from the centre of a region is given a weighting close to 0.
- Inverse distance between an estimation point for sub-region and the centre of that estimations sub-region.
- Inverse Manhattan distance between an estimation point for sub-region and the centre of that estimations sub-region.
- Inverse variance computed for a estimation point within sub-region by the GP for modelling that sub-region.

Other methods are possible. The first method is generally the more robust and much less susceptible to dramatic variation between adjacent sections.

## 5.3  Mesh Resolution

The modelling of the surface is done indirectly by a set of points, where each estimation point has a value representing which side of the surface that point likely lay on, see [5]. The Lewiner et al. [18] marching cubes algorithm is then used to find the surface that cuts between the two sections.

Mesh granularity affects accuracy of the surface produced. The finer the mesh the more accurate the surface. In this case the surface fidelity is affected both by the accuracy of the estimates, and also by the coarseness of the tessallation demarcating the estimates using marching cubes. As the mesh becomes coarser the accuracy declines and additional artifacts become noticeable, in particular what appears to be stair casing which can be seen in the centre sections of Fig. 4c, d. We can for instance visualize a mesh resolution of one full bench height, at this resolution surfaces that run at a shallow angle to the bench will run horizontally for a while then step up/down to the next bench and run horizontally for a while. This can be controlled by varying the resolution of the mesh (at the cost of computation). It is also worth noting that in a geological model the space is often turned into blocks of varying sizes. The minimum viable size for a block will be related to the physical characteristics of the diggers employed at a site and the amount of material movement that is normally seen during blasting. While improving mesh resolution may look nice, it will reach a point in which it has little/no practical value.

## 5.4  Model Evaluation

Results from the use of a global rotation model on a mineralised boundary and a stratified boundary are compared to boundaries generated using the original local rotation calculations (Fig. 4). The two different boundary types are modelled in two spatially different locations within the Pilbara region of Western Australia.

By having a global rotation model, the rotation transformations for local sub-regions are now spatially correlated. From Fig. 4 we can see that this has ensured a level of similarity between neighbouring sub-regions, less dramatic artifacts between neighbouring sub-regions, as can be seen when comparing Fig. 4c, d.

To compare model performance, we calculated reconciliation values for tonnes for two relevant portions of the mine. Reconciliation values are a comparison of what a particular model predicted vs what was extracted. In this case we produced three GP estimation models for the area. The first model (Fusion) was created using the boundaries produced by this work, including both the apriori data and the GP rotation. The second model (Warping) was created using boundaries produced by Bayesian surface warping which reduces inaccuracies in a modelled boundary with respect to new assay observations via displacement likelihood estimation [7]. The third model (Exploration) was created using the original exploration based boundary surfaces. The reconciliation values were calculated by comparing estimates from the

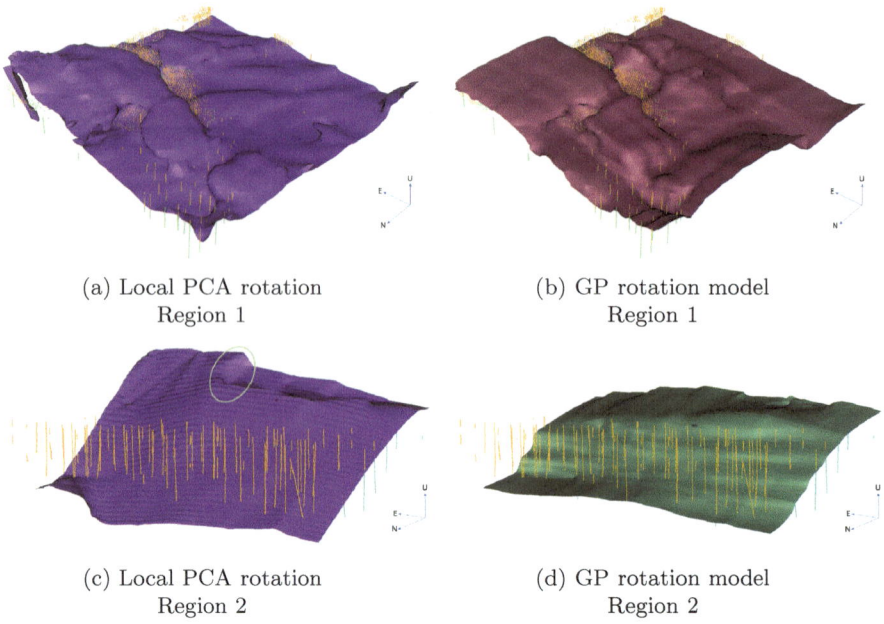

(a) Local PCA rotation
Region 1

(b) GP rotation model
Region 1

(c) Local PCA rotation
Region 2

(d) GP rotation model
Region 2

**Fig. 4** Comparison of boundary estimates from modelling pipelines that use two different methods for determining local region rotational transformations. This comparison was performed on two different underlying surfaces in two spatially different regions. Figures **a** and **c** presents a boundary estimate where each local regions rotation was calculated through PCA. Circled are regions where the surface looks 'step-like' as a result of neighbouring local regions having drastically different rotation functions. Figures **b** and **d** presents a boundary estimate with a trained GP model for inferring the rotation function across the global space. We can see that the boundary estimate is smoother and the 'steps' from **a** are not present

GP models to the values calculated for the same region using the production hole data. More details on the reconciliation procedure and differences between bench within and bench below prediction are illustrated in [7]. A value closer to 0 indicates a better prediction. At the bench within level, where the composition of the lowest bench containing blast hole data is predicted, the proposed model outperforms the other models on all grade block categories (Tables 1 and 2). When predicting on the bench below the available data, our model consistently outperforms the original model based on the exploration holes and has a comparable performance to the surface warping model (Tables 3 and 4).

## 6  Unassayed Production Holes

In an operational open-pit mine, production holes are holes drilled into mining benches in preparation for their blasting. Samples from production hole drillings are routinely collected for assaying [19], allowing for updated boundary models to

**Table 1**  Bench within reconciliation results for site 1

| Grade block | Tonnage | Mean absolute difference | | |
|---|---|---|---|---|
| | | Fusion | Warping | Exploration |
| High grade | 434196 | 0.023 | 0.036 | 0.100 |
| Low grade | 74865 | 0.010 | 0.029 | 0.118 |
| Waste | 963091 | 0.057 | 0.068 | 0.131 |

**Table 2**  Bench within reconciliation results for site 2

| Grade block | Tonnage | Mean absolute difference | | |
|---|---|---|---|---|
| | | Fusion | Warping | Exploration |
| High grade | 656117 | 0.014 | 0.018 | 0.056 |
| Low grade | 422379 | 0.031 | 0.033 | 0.054 |
| Waste | 10076 | 0.028 | 0.060 | 0.091 |

**Table 3**  Bench below reconciliation results for site 1

| Grade block | Tonnage | Mean absolute difference | | |
|---|---|---|---|---|
| | | Fusion | Warping | Exploration |
| High grade | 129831 | 0.009 | 0.006 | 0.043 |
| Low grade | 28820 | 0.062 | 0.060 | 0.173 |
| Waste | 534777 | 0.077 | 0.089 | 0.145 |

**Table 4**  Bench below reconciliation results for site 2

| Grade block | Tonnage | Mean absolute difference | | |
|---|---|---|---|---|
| | | Fusion | Warping | Exploration |
| High grade | 85307 | 0.063 | 0.062 | 0.079 |
| Low grade | 57334 | 0.032 | 0.032 | 0.049 |
| Waste | 861600 | 0.074 | 0.072 | 0.099 |

be produced from the new data [7, 8]. In mining scenarios where 'ore' and 'waste' are visually differentiable, production holes in 'waste' regions will not be assayed for temporal and financial reasons. This has a deleterious effect on boundary models generated through automated processes as the absence of observation data on the waste side of the ore/waste boundary will reduce the accuracy of the resulting boundary estimate.

It is therefore desirable to have some estimation method for determining whether an unassayed production hole is a waste hole omitted from the assaying process, or is unassayed for some other reason. Some other reasons for not assaying a production hole are: every $n$-th hole is assayed (for temporal and financial reasons), for quality control (i.e. the values in the assay are not considered correct), or the hole was drilled for blast control reasons, removing the need for assaying. The omission of only waste hole observations biases predictive models, resulting in poor model performance. This makes them non-ideal in deciding whether an unassayed production hole should be re-introduced into the boundary modelling process.

We therefore present a heuristic for the labelling of unassayed holes from the assayed production holes with 'ore' and 'waste' labels. This heuristic is only used for consideration as to whether an unassayed hole should have a 'waste' label. While we do not assume that all unassayed production holes are waste holes, we do assume that 'ore' regions—regions of interest—are sufficiently represented.

In constructing the heuristic to determine whether unassayed production holes should be labelled as 'waste', we assume that a spread of production holes (combination of assayed and unassayed) is provided, with the assayed production holes having an associated 'ore' or 'waste' label. From here, a series of criteria for labelling an unassayed production hole with a 'waste' label can be formulated. For this paper, the criteria were:

- The production hole in question must be sufficiently far from all 'ore' labelled production holes. The nearby presence of a 'waste' label should not prevent, or be required for, the labelling of a production hole as waste. In our scenario, the distance threshold was 10m.
- The production hole must not be one drilled for the purpose of blast control in the bench. This can be determined by ensuring that the hole is 'sufficiently' vertical and has a depth typical of production holes drilled on site. For our example, this distance was 9–12 m. Some mining operations also assign particular codes to production holes drilled for the purpose of blast control, which can also be incorporated into the heuristic.

## 6.1 Results

Figure 5a presents an area which has several production holes that we would like to assign a label to. For the purposes of demonstration, the production holes that we are attempting to label have been assayed and have a total assay percentage of 98–99%. These are production holes that have not been included in the original modelling process, but have sufficient data to be used in heuristic validation. When the proposed heuristic was applied, a total of 25 production holes were eligible to be labelled as a waste hole. Comparing these holes against their recorded assay values yielded a labelling accuracy of 100%. The newly labelled production holes can be seen in Fig. 5b.

We can see that there is a trade-off when setting heuristic parameters such as the minimum distance to a non-waste hole. Reducing this distance will mean that more unassayed holes will be labelled, but the likelihood of these production holes being incorrectly labelled increases. If the threshold in our presented example was reduced

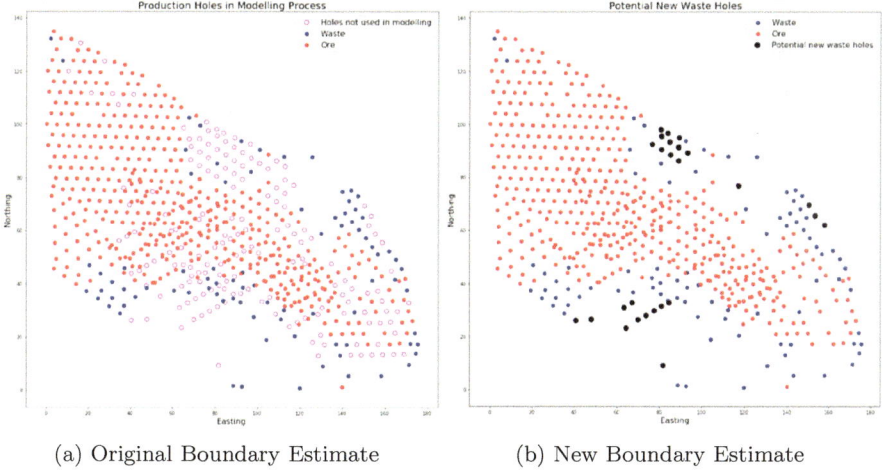

(a) Original Boundary Estimate          (b) New Boundary Estimate

**Fig. 5** Example of production holes. The red (ore) and blue (waste) dots represent the location of assayed production holes ready for inclusion in a modelling process. In **a** the hollow magenta dots represent unlabelled production holes, while in **b** the black dots represent unassayed production holes that will be assigned a waste label

to 9 m, 1 out of 34 holes would be incorrectly labelled (97.06% accuracy), while if the threshold was 8 m, 1 out of 44 holes would be incorrectly labelled (97.73% accuracy).

The boundary estimates resulting from the production hole data in Fig. 5 and exploration hole data is shown in Fig. 6. Both of the boundary estimates are shown with the augmented production hole data. When the newly labelled production holes are omitted from the modelling process (Fig. 6a), the boundary estimate passes under these holes (centre-right). When these holes are included in the modelling process (Fig. 6b), the boundary estimate has risen to put these holes in the waste region.

## 7 Discussion and Conclusions

The inclusion of a priori data and previously unlabelled waste production holes both provide a mechanism for incorporating expert domain knowledge into a boundary modelling process, improving boundary estimation models. The inclusion of a priori data also contributes to the generation of boundary estimates that better align with geological expectations in data sparse regions, and can prevent the generation of some artefacts, including false isosurfaces, in the boundary estimates. By using a global, continuous GP model to determine local sub-region rotation transformations for a stitched large-scale GP boundary modelling process, a further improvement in boundary estimates can be realised.

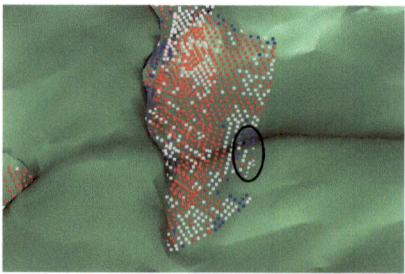
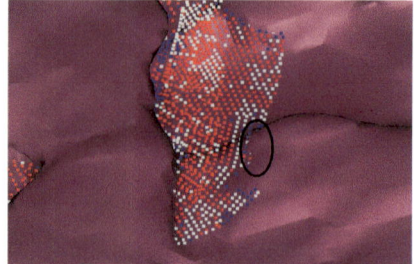

(a) Original Boundary Estimate            (b) New Boundary Estimate

**Fig. 6** Comparison of mineralisation boundary estimates for a grade block. The first figure shows the boundary estimated using the original production hole labelling, while the second figure shows a boundary estimated using the augmented production hole labels. In both images, the augmented production hole labels are shown (red = ore, blue = waste, grey = unlabelled). This allows us to see the change in the boundary estimate near the centre right of the data (circled), where the boundary estimate from the augmented data has lifted to cover the newly labelled holes

Although the contributions presented in this paper improved the boundary modelling process in the demonstrated regions, some modification of these steps would be required for application to a different region due to modelling parameters being scenario specific. This is something of interest that we hope to explore in the future. Further to this, there is room for exploration into alternate representations for surface normals for the generation of local sub-region rotation models. Some examples here include: (1) estimating surface normals from a Delaunay mapping based on exploration hole transition point data, (2) having a more complex inferencing method for each local sub-region, such as taking the average of the transformation functions from the corners of the sub-region, rather than just using the inferred rotation from the centre of the sub-region. As accurate modelling is not typically performed in waste regions, quantitative assessment of the proxy production hole labels was not possible in this study. In the future, we plan to work with industry to acquire suitable data for these metrics.

In conclusion, this paper has presented three application-based solutions to challenges in the automated generation and updating of boundary estimates: (1) The inclusion of a priori data, allowing for the incorporation of domain expertise into the boundary modelling process. (2) A heuristic for the labelling of unassayed production holes, improving boundary modelling accuracy. This further increases the incorporation of domain expertise into the modelling process through the augmentation of available data. (3) The incorporation of a global modelling process for the calculation of rotation transformations for local sub-regions that are used in the generation of a large-scale probabilistic boundary estimate. For the two sites considered in this work, the boundary fusion method improved the predicted tonnages for both bench within and bench below reconciliations when compared to the exploration based boundary surfaces. When compared to the surface warping model, boundary fusion had comparable results on the bench below and improved results on the bench within. These solutions allow for the generation of more accurate boundary estimations which in turn improves the fidelity of ore-grade estimation models. Demonstration of these solutions has been presented on both stratigraphic and mineralisation boundaries in the Hamersley Province of Western Australia.

# References

1. Everett, J.E.: Planning an iron ore mine: from exploration data to informed mining decisions. In: Proceedings of the Informing Science and Information Technology Education Conference, pp. 145–162. Informing Science Institute (2013)
2. Yang, L., Achtziger-Zupančič, P., Caers, J.: 3d modeling of large-scale geological structures by linear combinations of implicit functions: application to a large banded iron formation. Nat. Resour. Res. 1–25 (2021)
3. Laurent, G.: Iterative thickness regularization of stratigraphic layers in discrete implicit modeling. Math. Geosci. **48**(7), 811–833 (2016)
4. Neves, J., Pereira, M.J., Pacheco, N., Soares, A.: Updating mining resources with uncertain data. Math. Geosci. **51**(7), 905–924 (2019)
5. Ball, A., Silversides, K.L., Chlingaryan, A., Melkumyan, A.: Creating large scale probabilistic boundaries using gaussian processes. Expert Syst. Appl. **199**, 116959 (2022)
6. Melkumyan, A., Ramos, F.: Non-parametric Bayesian learning for resource estimation in the autonomous mine. In: 35TH APCOM Symposium Wollongong, Australia, pp. 209–215 (2011)
7. Leung, R., Lowe, A., Chlingaryan, A., Melkumyan, A., Zigman, J.: Bayesian surface warping approach for rectifying geological boundaries using displacement likelihood and evidence from geochemical assays. ACM Trans. Spat. Algorithms Syst. **8**(1) (2021)
8. Leung, R.: Modelling orebody structures: block merging algorithms and block model spatial restructuring strategies given mesh surfaces of geological boundaries. J. Spat. Inf. Sci. **21**, 137–174 (2020)
9. Thorne, W., Hagemann, S., Webb, A., Clout, J.: Banded Iron Formation-related Iron Ore Deposits of the Hammersley Province, Western Australia, vol. 15, pp. 197–221. Society of Economic Geologists, Westminster, CO, USA (2008)
10. Trendall, A.F., Blockley, J.G.: The iron formations of the Precambrian Hamersley group, Western Australia with special reference to the associated crocidolite. Bulletin **119**. Geological Survey of Western Australia, Perth (1970)
11. Dalstra, H.J., Rosiere, C.A.: Structural controls on high-grade iron ores hosted by banded iron formation: a global perspective. Rev. Econ. Geol. **15**, 73–106. Society of Economic Geologists, INC, Littleton, CO (2008)
12. Harmsworth, R.A., Kneeshaw, M., Morris, R.C., Robinson, C.J., Shrivastava, P.K.: BIF-Derived Iron Ores of the Hamersley Province, pp. 617–642. The Australasian Institution of Mining and Metallurgy, Melbourne (1990)
13. Rasmussen, C.E., Williams, C.K.I.: Gaussian Processes for Machine Learning. MIT Press, MA (2006)
14. Zaitouny, A., Small, M., Hill, J., Emelyanova, I., Clennell, M.B.: Fast automatic detection of geological boundaries from multivariate log data using recurrence. Comput. Geosci. **135**, 104362 (2020)
15. Larrondo, P., Deutsch, C.V.: Accounting for geological boundaries in geostatistical modeling of multiple rock types. In: Deutsch, C.V., Leuangthong, O., (ed.) Geostatistics, pp. 3–12. Springer, Dordrecht (2005)

16. Linsel, A., Wiesler, S., Haas, J., Bär, K., Hinderer, M.: Accounting for local geological variability in sequential simulations–concept and application. ISPRS Int. J. Geo-Inf. **9**(6), 409 (2020)
17. Pérez, C., Mariethoz, G., Ortiz, J.M.: Verifying the high-order consistency of training images with data for multiple-point geostatistics. Comput. Geosci. **70**, 190–205 (2014)
18. Lewiner, T., Lopes, H., Vieira, A.W., Tavares, G.: Efficient implementation of marching cubes' cases with topological guarantees. J. Graph. Tools **8**(2), 1–15 (2003)
19. McArthur, G., Jones, C., Murphy, M: Blasthole cone sampling experiments for iron ore flitch mining. In: Proceedings Sampling (2010)

# Appendix A
# Appendix: Short Abstracts

This appendix encompasses the short abstracts of all works presented in the 11th International Geostatistics Congress that were not included as extended abstracts or full articles.

## Theory

### Regularization and Deregularization of Unidimensional Covariance or Variograms[1]

Christian Lantuéjoul*, D. E. Bush, J. Stiefenhofer and M. L. Thurston
*christian.lantuejoul@mines-paristech.fr

Back to basics! Regularizing a point covariance or variogram on a segment results in a linear combination of not less than 9 terms. It is not so easy to understand how these terms are related. As a consequence, there is a risk of getting an erroneous result. Moreover, there is also a risk of not spotting it. In this presentation, we show how these terms are structured and provide a safe and effective procedure to derive them. An important application of this procedure is the deconvolution problem, that is the conversion of a regularized variogram to a point one. This problem is known to be an ill-posed one. It is sorted out using a Bayesian approach that produces a family of point variograms that are compatible with the initial regularized variogram.

---

[1] A modified and extended version of this work has been submitted to the Geostats 2021 Special Issue in the journal Mathematical Geosciences.

© The Editor(s) (if applicable) and The Author(s), 2023
S. A. Avalos Sotomayor et al. (eds.), *Geostatistics Toronto 2021*, Springer Proceedings in Earth and Environmental Sciences, https://doi.org/10.1007/978-3-031-19845-8

Subsequently, each point variogram can be regularized at any support, which results in a fully compatible family of variograms and cross-variograms at different supports.

**Keywords:** Regularization; Variograms; Cross-variograms

### Multivariate Cross-Validation and Measures of Accuracy and Precision[2]

U. Mueller*, K. G. van den Boogaart and R. Tolosana-Delgado
*u.mueller@ecu.edu.au*

Cross-validation and jack-knifing are established methods for validating the geostatistical model to be used in either estimation or simulation. The standard outputs, scatterplots of estimated against true values, (standardised) estimation error against estimates and error statistics, are suitable for the estimation/simulation of univariate data or for cases where a clear primary variable with one or more secondary variables is to be modelled. However, in the case of truly multivariate data (such as directional or compositional data), there is no hierarchy of variables, in that the entire regionalised vector needs to be modelled. Thus, geostatistical estimation and simulation in this case need to be treated as fully multivariate and any appraisal of the goodness of the geostatistical model needs to take this aspect into account. This concerns not only cross-validation or jack-knife approaches, but also accuracy and precision of simulations.

The direct assessment of local accuracy and precision was first discussed in the context of univariate geostatistical simulation by Deutsch at Wollongong 96. The assessment was based on the local distributions of simulated data at locations where the true value was known derived either from cross-validation or jack-knifing. In either approach the location of the true value relative to the mean of the local distribution was used assess the widths of the local distributions. For each probability $p$ and each location $u$, one considers, whether the value $v(u)$ is contained in the $p$-interval $\left[ F_u^{-1}\left(\frac{1-p}{2}\right), F_u^{-1}\left(\frac{1+p}{2}\right) \right]$. The function $F_u^{-1}$ denotes the inverse local CDF. The simulation algorithm then was termed accurate, if the proportion $\xi(p)$ of locations falling into to the $p$-interval exceeds $p$, meaning that the distribution is "wide enough". Precision is then defined in terms of the difference between $\xi(p)$ and $p$ and one would like these to quantities to be close. A useful mechanism for appraising the accuracy of the simulation is a plot of $\xi(p)$ against $p$.

In this contribution we consider the evaluation of the suitability of a geostatistical simulation model in the compositional framework, i.e. where each variable is positive and its values inform of the relative abundance of a certain component forming the system. Specifically, we will work under the assumption of additive logistic normality and derive validation measures for this scenario. We adopt the principle of working in coordinates, allowing to define the multivariate random function as Gaussian in

---

[2] A modified and extended version of this work has been submitted to the Geostats 2021 Special Issue in the journal Mathematical Geosciences.

log-ratio transformed scores. At each sample location $u$ the log-ratio cokriging estimates and error covariance matrix define a normal distribution $F_u$ (conditioned on the sample data) with expected value $z_k^*(u)$ and cokriging error covariance matrix $\Sigma_k(u_\alpha)$. From the estimates and error covariance matrix we may construct univariate measures (treating each variable as the primary variable at a time) but also vector valued errors. A suitable error measure is then the Aitchison Mahalanobis norm of the difference between estimates and true values, which has a $\chi^2(D-1)$ distribution under the hypothesis of additive logistic normality of a $D$-component random function. Similarly, the indicator variable used for computing accuracy statistics needs to be redefined in terms of the Mahalanobis norm. Here the indicator variable measures whether the true value is contained within a ball of radius $p$ of the estimated value with respect to the Aitchison Mahalanobis norm. As in the univariate case the function $\xi(p)$ then defines the proportion of locations for which the true composition is contained in the ball of radius $p$ about the estimate. We evaluate the usefulness of this approach via the validation of a simulation model for a high-dimensional regionalised composition.

**Keywords:** Simulation cross-validation; Compositional data; Accuracy: Precision

### High-Performance Grid-Less Geostatistics with Distributed Computing

Alexandre Boucher* and Pericles Machado
*aboucher@ar2tech.com

Theoretically, variogram-based geostatistical estimations and simulations are independent of a grid. The sample locations, their values, the variogram or cross-variogram models and possibly a trend model are all that is required. All other parameters are specific to the implementation of kriging estimation and geostatistical simulation algorithms. Thus, when using algorithms that are non-sequential and grid-independent, such as kriging and turning bands simulation, there is an implicit representation of a spatial phenomenon. Any numerical representation on a grid is simply an on-demand sampling of that implicit model, be it a mineral resource block model or a flow simulation grid. This grid-less approach of spatial modelling is particularly powerful when the geostatistical algorithm is available on a high-performance platform, such as cloud-based distributed computing. At that point, one can quickly and consistently create numerical views of the models on different grid types and supports. For instance, values from an implicit model generated by a set of turning bands can be consistently extracted with a coarse regular grid, a heterogeneous unstructured grid or a set of virtual drill holes. Each of these representations has different scale of support and cell configuration but sample the same implicit model generated by the parameterized algorithm, therefore enforcing consistency with one another. Finally, with algorithms designed to run on high-performance computing platforms, the grid-less approach frees the modeler from predefined grid topologies and allows adjusting the model numerical representations to the task at hand, such as passing arbitrary cross-sections to a visualization engine.

**Keywords:** Grid-less geostatistics; Simulation; Estimation; Distributed computing

### Sequential Simulation of a Boolean Model

Alan Troncoso*, Xavier Freulon and Christian Lantuéjoul
*alan.troncoso@mines-paristech.fr

Modeling the heterogeneities within a geological unit is a key issue when assessing the dynamic response of an underground reservoir. Typical applications include the prediction of oil recovery, the estimation of groundwater contaminations or the integrity evaluation of a $CO_2$ storage. Many approaches have been developed to cope with such a task, e.g., gaussian, object and process-based models. In these models, there is a trade-off between global realism and local consistency with observations. For instance, the plurigaussian simulations easily honor the facies observed along wells but fail to reproduce complex geological relationships, whereas object based or process-based models can better reproduce the conceptual geological model but may fail to be conditioned to observations. In this study we propose to apply a divide and conquer strategy to produce conditional simulations of object-based models: the approach considered consists in decomposing this complex simulation problem into a series of simpler ones. To do so, Sequential Monte-Carlo techniques or particle filtering have been adapted to the geostatistical context. This sequential approach is illustrated with the Boolean model. When subject to pore and grain conditions, this model can be expressed as the union of two independent random sets: a first Boolean model made of all objects subject to grain conditions, and a second Boolean model subject only to pore conditions. This decomposition greatly simplifies the integration of the constraints. The standard iterative simulation method based on a birth-and-death process on the random objects is compared with this alternate sequential approach. Two synthetic cases are used to illustrate both methodologies: a first one in two dimensions that represents a vertical section of a channelized reservoir and a second one in three dimensions that represents the reservoirs. We expect that this sequential approach will be soon applied to more complex models such as meandering channels.

**Keywords:** Sequential simulation; Boolean model; Reservoir modeling

### A Non-stationary Linear Model of Coregionalization

Alvaro Riquelme* and Julian M. Ortiz
*alvaro.riquelme@queensu.ca

Multivariate modeling for natural resource characterization requires an underlying model of coregionalization. Among different tools for simultaneous modeling of variables, the linear model of coregionalization (LMC) is the most used as it can combine a large number of data types measured at different locations and different data support into the same framework, remaining as a useful and mathematically flexible tool. The LMC considers k co-regionalized variables as a linear mixture of nst independent factors, at each location of the stationary domain. A natural extension for the LMC is to alter the linear mixture, which is assumed fixed on the domain. This linear mixture can be made locally varying according to the local strength in the dependency of the co-regionalized variables, leading to a locally varying linear model of coregionalization (LVLMC). The main challenge, once this relaxation on the LMC is assumed, is to solve appropriately the interpolation of the different known correlation matrices throughout the domain, in a reliable and coherent fashion. Correlation matrices belong to the family of symmetric positive definite (SPD) matrices, which in turn forms a cone shape Riemannian manifold. Building upon earlier studies that have shown that a Riemannian framework is appropriate to address the challenge of interpolation between correlation matrices, a brief overview of the geometric properties of SPD manifold is introduced, together with the properties that correlation matrices inherit from this SPD manifold. The present work adopts this non-Euclidean framework to achieve our objective by locally averaging and interpolating the correlations between the variables, retaining the intrinsic geometry of correlation matrices and using existing methods that are computationally efficient.

**Keywords:** Linear model of coregionalization; Eigen-decomposition; Geodesics; Riemannian manifold; Symmetric positive definite

## Gibbs Sampling and Successive Over-Relaxation for Simulating Gaussian Random Vectors

Daisy Arroyo* and Xavier Emery
*darroyof@udec.cl

The simulation of Gaussian random vectors and random fields arises in many disciplines of the natural sciences. This work presents two iterative algorithms aimed at simulating a Gaussian random vector Y with zero mean and given variance-covariance matrix C. This vector can correspond to the restriction at finitely many locations of a Gaussian random field, without any restriction on its spatial correlation structure (stationary or not, uni-variate or multivariate) or on the space (Euclidean or not) in which it is defined. The first algorithm pertains to non-conditional simulation and is a variant of the Gibbs sampler, in which a relaxation parameter is introduced in order to improve the rate of convergence to the desired Gaussian random vector. The second algorithm aims at conditioning the simulation to a set of hard data and is based on the method of successive over-relaxation. The novelty of both algorithms is that they are formulated in terms of the dual random vector $X = C^{-1}Y$ and do not require pivoting, inverting or square rooting the variance-covariance matrix C, hence they are applicable to simulate large random vectors or to condition the simulation to a large data set. Numerical experiments are performed to check the accuracy of the algorithms and to determine the relaxation parameters that optimize the rates of convergence, based on the deviation between the expectation vector and variance-covariance matrix of the simulated Gaussian random vector and the target expectation vector and variance-covariance matrix C, in both the non-conditional and conditional cases.

**Keywords:** Gibbs sampler; Non-conditional simulation; Successive over-realization; Conditional simulation

## Quantification of the Space of Uncertainty and Its Applications in Geostatistical Modelling

Maryam Hadavand* and Clayton V. Deutsch
*mhadavand@slb.com

The current practice of geostatistical modelling of categorical variables considers K = 2–7 categories and N = $10^6$–$10^8$ locations. The maximum possible space of uncertainty being $K^N$ is inconceivably large and cannot be understood from a practical perspective. This space becomes much smaller and calculable in presence of unequal category proportions, spatial correlation, and conditioning data. A general framework is presented to calculate the size of the space of uncertainty. This is very interesting to appreciate what would be required to understand the space of uncertainty. This becomes practically relevant when rejection sampling approaches are being used to condition geostatistical models as in the case of stochastic inversion. The size of the space of uncertainty is shown to be the product of exponential entropy values. This is corroborated from information theory, but the application of this in presence of spatial correlation and conditioning data is new. An implementation of the calculation is used to demonstrate the size of uncertainty for different cases. Practical consequences of this calculation are discussed.

**Keywords:** Geostatistical approaches; Multiple realization

## Bayesian Inversion into Soil Types with Kernel-Likelihood Models

Selamawit Moja* and Henning Omre
*selamawitserka.get@gmail.com

During construction of the off-shore wind mill foundations, the geotechnical engineers must study the soil properties in the sub-surface. These soil characteristics can usually not be directly observed, but related variables can be observed from well logs along the vertical profile. In the previous study (Moja et al. 2018), we constructed a prediction rule for the sub-surface facies characteristics from well-log observations. This rule is based on a non-stationary prior Markov chain model with a Gaussian likelihood model on factorial form. However, the Gaussian likelihood model does not capture the bimodal nature of our CPT data. The non-stationary prior models are formulated, hence, the time independent transition matrices and varying marginal probabilities are obtained. In the current study, we define a non-parametric kernel model in order to capture the bimodal nature of the observations which offers an alternative to traditional parametric models. The likelihood model for the observations is assumed to be in factorial form and they are assessed from a calibration well by kernel estimators. The prior Markov model are defined to be either a traditional stationary Markov chain or a trend Markov chain. For prediction of the sub-surface layer, both circular uniform kernel and Gaussian kernel likelihood model are defined and evaluated. The model parameter band width in the kernel likelihood is estimated by a cross-validation pseudo-likelihood estimation criterion. The methodology is demonstrated on one case study for offshore siting of wind-mills. The result from the current study provides improvement in the prediction of the subsurface profile. The sensitivity of the likelihood model to the choice of kernel band width is also explored. Thereafter, an estimate of the optimal band width based on the maximum cross-validation pseudo-likelihood criterion is obtained, and the corresponding likelihood models and posterior pdfs are presented. We conclude that a suitable choice of kernel likelihood model is of at most importance, and that using a trend Markov prior model improve the predictions even more.

**Keywords:** Sub-surface geotechnical prediction; Circular uniform kernel; Gaussian kernel; Kernel-likelihood; Bayesian inversion

### Generalization Error of Learning Models Under Covariate Shift and Spatial Correlation

Julio Hoffimann*, Maciel Zortea, Breno W. S. R. de Carvalho and Bianca Zadrozny
*julio.hoffimann@gmail.com

Statistical learning (a.k.a. machine learning) models are prone to overfitting: the condition that the empirical risk is much greater in an unseen dataset than in the dataset used for learning the model. If not assessed nor controlled, overfitting can invalidate the application of learned models in practice, particularly if model's results are to be used for making decisions involving natural resources. Assessing overfitting, or more broadly generalization error, is especially important when dealing with models of great expressivity (e.g. neural networks with multiple layers) since these models can often be fine-tuned to memorize datasets. Statistical learning theory provides methods for assessing generalization error and the literature is vast on this theme. However, most existing methods do not take into account the unique challenges of performing statistical learning in spatial set-tings. In particular, it is well known that model errors cannot be assumed to be independent and identically distributed in spatial data due to spatial correlation. Moreover, spatial trends in the data lead to covariate shifts between the domain where the model was trained and the domain where it will be applied, and this issue invalidates the use of classical cross-validation approaches based on random splits of the data. In this work, we propose a method for estimating generalization error of learning models under covariate shift and spatial correlation. By creating synthetic data with known spatial distribution, we com-pare our method with prior art for increasing shifts and correlation lengths.

**Keywords:** Machine learning; Statistical learning; Generalization error

## A Training Image Free High-Order Simulation Framework Based on Statistical Learning[3]

Lingqing Yao* and Roussos Dimitrakopoulos
*lingqing.yao@mail.mcgill.ca

Multi-point simulation and high-order simulation methods have been pro-posed to overcome the limitation of the traditional second-order geostatistical simulation methods in reproducing the complex spatial patterns. However, their applications are limited due to their reliance on a training image. The present work proposes a training-image-free high-order simulation framework using a statistical learning approach. The statistical learning mechanism aims at matching the high-order spatial statistics of the generated realizations to those of the available sample data. This learning is accomplished by embedding the training data extracted from the original samples through a spatial template into a newly designed kernel Hilbert space. Specifically, the conditioning data in the simulation initiate a so-called data event associated with a certain spatial template and the replicates of the data event with the same geometry configuration are extracted from the samples and utilized as the training data. A spatial Legendre moment kernel is proposed to construct the kernel Hilbert space so that the high-order spatial statistics of the original data space are carried to the new elements in the kernel space after feature mapping. Minimizing the distance of the elements corresponding to the target probabilistic model and the empirical statistical model in the kernel space leads to reproducing of high-order spa-tial statistics of the sample data, which amounts to solving a quadratic programming problem. To address the major challenge of lacking fully matched replicates of the data event in the case that a training image is not present, an approach of aggregating statistics in interrelated kernel subspaces is proposed herein to simultaneously utilize the high-order spatial statistics from the partially matched replicates in the learning process. Case studies show that the proposed method reproduces the spatial patterns of the data and is suitable for practical applications without a training image

**Keywords:** Multiple-point statistics; Training image; Conditional simulation

---

[3] An early version of this work has been published in the journal Mathematical Geosciences: Yao, L., Dimitrakopoulos, R. & Gamache, M. Training Image Free High-Order Stochastic Simulation Based on Aggregated Kernel Statistics. Mathematical Geosciences 53, 1469–1489 (2021). https://doi.org/10.1007/s11004-021-09923-3.

**Inequality Data in Multiple-Point Statistics Simulation**

Julien Straubhaar* and Philippe Renard
*julien.straubhaar@unine.ch

Multiple-point statistics (MPS) simulation is an efficient and flexible frame-work for modeling complex heterogeneous systems with a high realism. MPS simulations reproduce the spatial statistics given in a training data set while honoring conditioning data—known values at given locations. MPS can account for different types of non-stationarity: changes of spatial structures, local orientation, local proportions, etc.

Inequality data, consisting in one inequality or a target interval at a set of given locations, arises often in practice. For example, concentration data in a sample may be below a detection limit, or the elevation of the base or top of a geological formation maybe known to be below or above a certain altitude in a spatial domain. To handle such constraints in MPS simulations, a rejection approach, consisting in simulating without accounting for the inequalities and keeping the realizations honoring them, can be a solution when these data are sufficiently sparse, but it becomes inefficient or even intractable in the presence of a dense inequality data set.

In this work, we present an MPS method based on the Direct Sampling strategy and implemented in the DeeSse code that is able to account for any inequality data set. It consists in adapting the computation of the pattern distance (dissimilarity) between the pattern centered at the current cell in the simulation grid and the patterns scanned in the training data set. This new distance account for interval of values, instead of exact values, in a part of the nodes retrieved from the simulation. The technique is illustrated with an example where we simulate topography in a two-dimensional area.

**Keywords:** Multiple-point statistics; Conditional simulation; Inequality constraints; DeeSse

# Petroleum

### Electrofacies Classification to Improve Conditioning Data for Integrated Geological Modeling

David Garner*
*david@terra-mod.com

Many hydrocarbon reservoir studies rely on detailed geological modeling as a useful tool for forecasting flow behaviour and planning development scenarios. Underlying the integrated models is a geological conceptual model including a form of facies description and pattern of succession. A key impact on hydrocarbon reservoir studies is a rigorous strategy around developing facies for modeling purposes. The current industry best practice for modeling reservoir heterogeneities related to flow is to apply a hierarchical workflow of simulation of a facies variable first, followed by property simulations within each modeled facies. Model facies are a categorical variable, typically input as facies indicator logs, used for hard conditioning in the geomodelling process, a "truth" variable. Yet, the facies variable delivered to modelers can often have ambiguity and imprecision due to the sparse subsurface sampling of information and from both visual and numerical procedures used to determine them. Classically, the given facies are a visual interpretation of the face of a rock sample consistent with the current geological concepts. These concepts may influence many aspects of the geological modeling procedures. For geostatistical modeling purposes, the input facies categories are indicator variables assumed to each represent statistically stationary domains of reservoir properties within the region of interest, the study area. In practice, the stationarity assumption is violated requiring auxiliary modeling steps to account for non-stationarities related to trends in facies deposition and/or properties such as compaction of porosity or fluid gradients. Fluid distributions as well as flow and mechanical properties are dependent on the characterization by each facies to effectively model useful changes in the reservoir. Accounting for realistic physical behavior, such as percolation and capillarity, when distributing properties by facies ensures reasonable physical responses in fluid flow models and in direct forecasting methods. Establishing early the petrophysical distinctness of facies variables input to models through electrofacies classification methods can mitigate common detrimental issues that ultimately degrade the fidelity of model verisimilitude. Electrofacies modeling provides a robust and dedicated framework of methods to impose consistency on facies logs delivered for modeling, thus enhancing the capabilities for effective integration of multi-scale data for reservoir modeling.

The electrofacies classification processes typically apply multivariate statistics using wireline logs and visual core or image description for training sets. Electrofacies, because they are mainly derived from petrophysical curve responses, are close to being lithological predictions, like lithofacies with distinct rock property ranges and trends. The classification of lithofacies involves various approaches. Visual methods are well established combining rock fabric, pore space and petrophysics (e.g. Lucia 1995) within a working concept model. These usually include detailed description of depositional and diagenetic processes from outcrops, core samples and/or image log data. Petrofacies classification involves defining rules-based petrophysical categories, e.g. using log cutoffs or manually defined regions, a somewhat useful method in lower dimensions of two or possibly three variables. The practical advantage of electrofacies is consistency provided by statistically combining the visual geological classifications with a suite of petrophysical log data. The result, which is beneficial to subsequent geomodelling processes, is to enforce the distinct lithological characteristics at the log curve scale (Garner et al. 2014). This provides flexibility and options for handling the multiscale data integration to start and during geomodelling.

Visual facies may be used directly for modeling or as a part of a training set (visual facies and well logs) for numerical classification methods. Arguably, visual facies are a cognitively biased data type due to their origin. In practice there are errors in interpretation and in the core, well log and image data used for interpretation. A brief discussion of five assumptions underlying a linear discriminant analysis provides practical guidance on the need for checking, cleaning and improving the usefulness of facies inputs to models. To establish rules for improving classification training sets, we discuss assumptions for a parametric method, linear discriminant analysis, as described by Davis (1986) to highlight common issues. The goal is to improve the classification results whether using parametric or non-parametric methods. These five parametric assumptions are all violated to some degree by the training sets (from Garner 2019):

- The observations in each class were randomly chosen. (Observed facies are not random samples. They are spatially biased as is the natural variability of depositional successions and observed stratigraphic sequences).
- The probability of an unknown observation belonging to each class is equal (Facies proportions are not equal in nature. Facies proportions are initially defined and honoured during modeling. The consequence is we may adjust the final electrofacies assignment model with weights to reasonably honour inputs).
- Variables are normally distributed within each class. (By-facies distributions of log variables have various shapes in the hyper-space, especially with depositional facies concepts).
- The variance-covariance matrices of the classes are equal in size. (The multivariate spread of properties for a given facies may be narrow or wide depending on lithological characteristics. That is the actual category samples are more or less densely clustered).
- None of the observations used to calculate the function were misclassified. (Facies and logs have imprecision with many sources of error leading to erroneous petrophysical statistics, e.g. depth shifts, interpretive scale used by the geologist, bed boundary overlap between data types, petrophysical log normalization, interpretive ambiguities are among the common sources of potential "errors" and uncertainty).

Discriminant analysis, described by Davis (1986) albeit useful to understand for rules to guide cleaning of a training data set, is a parametric method applicable to simply organized data distributions, separable clusters and is not optimal for typically complex geological facies log data distributions. When using visual facies and well logs as the training sets for supervised electrofacies classifications, non-parametric methods (Nivlet et al. 2002; Ye and Rabillier 2000) tend to be most effective given the varied sizes, shapes and overlap of the geologically derived visual facies in the hyper-space, the multivariate distributions. The non-parametric methods use the data as given to establish probabilities and likelihoods of membership. Data clean up following these quasi-rules can improve the consistency of electro-facies classification.

To establish the workflow, thorough training set preparation is imperative for electrofacies methods to succeed. The visual facies are regularly defined at a different resolution and sampling scale than well logs, are prone to systematic errors, and have overlapping petrophysical property distributions. Cleaning involves inspecting and trimming input facies based on the outlier tails of the distributions for each log parameter. These represent measurement error and ambiguous information. Paradoxically, cleaning the training set entails interpretive judgement, a cognitive bias, and will alter the statistical measures used to check the results, e.g. during validation, increasing the percentage of correct assignments and changing reference facies proportions. However, once the training set is deemed cleaned, subsequent electrofacies parameter options may be compared consistently to one another. We will discuss and show the choice of curves to be assessed physically and statistically in a step-wise manner.

When enough wells are available with core facies, withholding a set for blind tests and for model validation can be pursued in the workflow. Validation can illustrate the robustness of methods and consistency with the underlying geological concepts. The final electrofacies logs will be judged by correct assignment rates, proportions being honoured, honouring of distinct statistical properties, and will be used to verify reasonable observance of the desired geological patterns. Assignment errors tend to be a reclassification to an adjacent quality facies, practically aiding lithological consistency for future heterogeneity modeling. Thus, the process is a guided one and not statistically unbiased. Examples from a few fields will be shown along with aspects of the workflows (Garner et al. 2009; Garner et al. 2014; Martinius et al. 2017). These examples will illustrate practical decisions, technical limitations, and options for modifications of methods as opportunities to further mature the technologies. Electrofacies modeling workflow steps are not widely established in the industry practice. There is a lack of best practice guidance, dissemination across technical disciplines and training. Perceived workflow complexity without these rules may have lead to misuse or sub-optimal application holding back use of this technology.

Visually interpreted facies must be checked for petrophysical consistency, i.e. the distinctness of petrophysical distributions, which is never guaranteed. Application of electrofacies methods, a multivariate classification can improve consistency for multi-scale data integration and is beneficial for the hierarchical steps within geomodeling workflows (Martinius et al. 2017). The electrofacies practice can be treated as an interpretive tool, a guided machine learning process, to obtain improved facies logs for input to reservoir models. The industry practices around preparing facies logs for modeling are diverse, field specific, and can benefit from the application of electrofacies classification workflows and the associated thought processes.

**Keywords:** Facies modeling; Electrofacies; Discriminant analysis; Non-parametric methods; Petrophysical logs; Data cleaning

**References:**

- Davis, J. [1986] Statistics and Data Analysis in Geology. 2nd Edition, John Wiley & Sons, New York, 646 pages.
- Garner, D., Lagisquet, A., Hosseini, A., Khademi, K., Jablonski, B., Strobl, R., Fustic, M. and Martinius, A. [2014] The Quest for innovative technology solutions for in-situ development of challenging oil sands reservoirs in Alberta. 2014 World Heavy Oil Congress, WHOC14-139.
- Garner, D., Woo, A., and Broughton, P., [2009] Applications of 1D Electro-Facies Modeling (abstract), CSPG Annual Convention, Calgary, May 10–13.

- Lucia, F. J., [1995] Rock-fabric/petrophysical classification of carbonate pore space for reservoir characterization. AAPG Bulletin, 79, 1275–1300.
- Martinius, A.W., Fustic, M., Garner, D.L., Jablonski, B.V.J., Strobl, R.S., MacEachern, J.A. and Dashtgard, S.E. [2017] Reservoir characterization and multiscale heterogeneity modeling of inclined heterolithic strata for bitumen-production forecasting, McMurray Formation, Corner, Alberta, Canada, Marine and Petroleum Geology, 82, 336–361.
- Nivlet, P., Fournier, F. and Royer, J.J. [2002] A new nonparametric discriminant analysis algorithm accounting for bounded data errors. Mathematical Geology, 34, 223–246.
- Ye, S. and Rabiller, P. [2000] A New Tool for Electro-facies Analysis: Multiresolution Graph-based Clustering. SPWLA 41st Annual Logging Symposium, June 4–7.

## Stochastic Pix2Pix Method for Conditional and Hierarchical Deepwater Reservoir Modeling

Wen Pan*, Honggeun Jo, Javier E. Santos, Carlos Torres-Verdín and Michael J. Pyrcz
*wenpan@utexas.edu

Unconfined Lobe depositional system reservoirs are one of the most common targets in deepwater oil field exploration and production. Seismic data integration is essential for obtaining accurate stochastic spatial property realizations. However, geological heterogeneity below seismic imaging resolution may control vertical and horizontal connectivity of the reservoir, hence affect oil production during development. Surface-based methods are commonly used for modeling these hierarchical structures but conditioning the models to well logs and geological horizons identified from seismic amplitude data is still difficult and time-consuming. Current geostatistical algorithms such as variogram- and multiple point-based simulation methods can easily be conditioned to well data and trends informed from seismic data at and above seismic resolution. Yet, such models are limited in their ability to reproduce essential heterogeneities below seismic resolution, including hierarchical structures and trends within lobes.

To solve the above problems, we develop a new machine learning algorithm, Stochastic Pix2Pix, to perform conditional stochastic subsurface modeling. This method extracts patterns from training models at different scales and stochastically combines them to generate diverse conditional realizations. We validate it using synthetic deepwater lobe reservoir modeling processes where (1) training models are efficiently generated with surface-based procedures, (2) seismic data are modeled as known horizons, and (3) wells with known properties are randomly placed in the reservoir. The obtained realizations success-fully match both the properties at well locations and the seismic horizons.

We successfully construct diverse 3D reservoir models conditioned to the well-log and seismic interpretations. Reproduction of heterogeneity from the training models is shown to be accurate with measures such as Lorenz coefficient and a new raster-based compensational index. In addition, model parameterization provided by this algorithm greatly accelerates the history matching process.

**Keywords:** Reservoir modeling; Pix2Pix; Data integration

## Managing Sparse Data and Missing Values in Unconventional Reservoirs: Classical Solutions in an Analytics-Driven Digital Tsunami

Jeffrey Yarus* and Melanie Adelman
*jmy41@case.edu

Sparsely available well and seismic data coupled with missing values due to incomplete measurements, measurement failure or error, are common challenges when building reservoir models. This is particularly true when modeling unconventional resource reservoirs where these specific challenges often go hand-in-hand. Unlike conventional reservoirs where vertical wells are the norm, pad-drilling coupled with horizontal wells are common practice in unconventional reservoirs. While this practice provides local increased well density, it does not necessarily imply borehole evaluation from logging tools, cores, or seismic acquisition. In fact, these important data are simply sparse. There are a variety of reasons for this including a confidence in local experience and more practically, a need to manage costs. As a consequence, this practice drives models to be more deterministic. They lack the benefit of understanding the level of uncertainty and provide little quantitative information on future well placement and completion strategy.

To further complicate the challenge, available existing data from both local vertical and horizontal wells are subject to missing values. Missing values can be corrected by petrophysicists who carefully analyze logs, making the appropriate corrections and interpretations. However, these modifications are time consuming, and in studies involving log data from many sources, there can be no guarantee that petrophysical domain expertise has been applied uniformly and appropriately across the entire data set. As a consequence, not all wells are properly scrutinized for a given study, and these important details often go unnoticed by modelers. This is particularly true for missing values.

Understanding how to build unconventional reservoir models in the presence of sparse data and missing values is critical, particularly as the industry moves toward high-performance distributed computing, automation, and machine learning. This study com-pares a variety of statistical and geostatistical methods for managing sparse data and mitigating the occurrence of missing values. It demonstrates the strengths, weaknesses, and synergies of both geostatistics and classical statistics with respect to the stated challenges, and offers cautious optimism with respect to the popular data science intervention and the current analytics-driven digital tsunami we are now experiencing.

**Keywords:** Sparse data; Unconventional reservoirs; Petrophysical logs
**Machine Learning Assisted History Matching for a Deepwater Lobe System**

Honggeun Jo*, Javier E. Santos, Wen Pan and Michael J. Pyrcz
*honggeun.jo@utexas.edu

Since the 1980s, depositional lobe systems have become important reservoir targets in the passive continental margin such as the Atlantic coast and the Gulf of Mexico. However, high exploration costs and complicated geologic structures challenge the reservoir characterization and modeling workflows. Even though seismic inversion and geologic interpretation is widely used to map the rock facies spatial distributions and their associated petrophysical properties, there is important sub-seismic resolution heterogeneity and geologic features that cannot be fully resolved. In the absence of sufficient high-resolution data, rule-based modeling has been applied to generate geostatistical stochastic realizations that quantify the uncertainty.

The stratigraphic rule-based reservoir models approximate sedimentary dynamics to generate realistic spatial distributions of petrophysical properties for reservoir forecasting and to support development decision making. A few intuitive rules combined with the sequential placement of surfaces bounding reservoir units render realistic reservoir hetero-geneity, continuity, and spatial organization to petrophysical property distributions that are difficult to obtain using conventional geostatistical pixel- and object-based subsurface models. However, as rule-based models incorporate the conceptual and qualitative information such as temporal deposition sequence and consequent compensational stacking patterns, integrating quantitative information such as production history data into the rule-based model has been one of the remaining obstacles to broad application.

This study proposes a machine learning assisted history matching workflow for rule-based models. First, multiple rule-based models are generated as training data for a Generative Adversarial Network (GAN). Then, the trained GAN is inspected both visually and statistically to check if it learned the primary geological features (e.g., depositional element geometries and hierarchical trends) in rule-based models and its realizations reproduce these features with a reasonable space of uncertainty. The successfully trained GAN enables exploration of the latent reservoir manifold to generate an ensemble of models. The initial ensemble models are fed to reservoir simulation to forecast production. Ensemble Kalman filter tunes the ensembles by minimizing the misfits between models' prediction and the given production observation.

This workflow results in a suite of reservoir models that honor both realistic geological heterogeneity and production history and provide a reasonable uncertainty model for reservoir forecasts. Moreover, this proposed workflow is computationally efficient as the entire process can be completed within a manageable time on a typical work station. The flexibility of this methodology of combining rule-based with generative machine learning models allows this workflow to be expanded to various depositional systems.

**Keywords:** Reservoir modeling; Generative adversarial network; Machine learning; History matching

**Single-Loop Geostatistical Seismic Inversion for Facies Prediction Combining Multiple-Point Geostatistical Simulation and Probability Fields Update**
Leonardo Azevedo* and Dario Grana
*leonardo.azevedo@tecnico.ulisboa.pt

Reliable subsurface modelling demands the simultaneous prediction of the spatial distribution of discrete and continuous properties, such as facies and rock properties. These models can be obtained by inverting seismic data, usually in a sequential two-step approach in the discrete or continuous domain. The optimization is generally performed in one of the two domains, discrete or continuous. Hence, the properties in the other domain are often simulated conditionally to the optimized variable but independently of the data mismatch.

In this work we propose a global iterative geostatistical seismic inversion method that couples stochastic sequential simulation and multiple-point geostatistical simulation as model generation tools of the continuous and discrete properties, respectively. The mismatch between synthetic and observed seismic data is used to update simultaneously discrete and continuous properties, ensuring a convergence of the iterative procedure in the facies and rock property domains.

The proposed method can be summarized in the following sequence of steps: (i) simulation of a set of facies models with multiple-point geostatistical simulation; (ii) simulation of a set of rock property models (e.g. porosity) with stochastic sequential simulation conditioned to the facies models simulated in (i); (iii) calculation of the elastic properties from the rock property models using a pre-calibrated rock physics model; (iv) comparison of the synthetic seismic models with the observed seismic on a trace-by-trace 2 basis; (v) selection of the facies and rock property samples that ensure the lowest mismatch values; (vi) updating of the probability fields in the multiple-point geostatistical simulation and conditioning of the stochastic sequential co-simulation of the continuous rock properties based on data mismatch. These steps are iterated until convergence.

The method is illustrated in synthetic and real case studies showing the ability to converge towards the true solution.

**Keywords:** Seismic inversion; Sequential simulation; Multiple-point statistics; Conditional simulation

## Subsurface Geological Modeling by Coupling Generative Adversarial Networks with Geostatistical Seismic Inversion

Leonardo Azevedo*, Arthur Santos and Gustavo Paneiro
*leonardo.azevedo@tecnico.ulisboa.pt

A central step in the geo-modelling workflow is the generation of a three-dimensional numerical model of the subsurface geological properties from a set of indirect measurements (e.g. seismic data). However, the relationship between model and data is highly non-linear and the prediction of the spatial distribution of the subsurface rock properties, which involves solving an inverse problem, is a challenging problem to address.

The simultaneous inversion of seismic data for facies and continuous properties is normally done using two outlines of a sequential approach. We can start by inverting seismic for facies from where the petrophysical properties are usually perturbed conditioned to the facies model. In this case, the update of the continuous properties is detached from the data mismatch. In the second outline, the seismic is inverted for petrophysical properties and facies are generated by classification after the model generation and perturbation and therefore not explicitly included in the stochastic optimization procedure.

The simultaneous inversion of continuous and discrete properties has been addressed with geostatistical inversion methods (Doyen 1988; Haas and Dubrule 1994; Mosegaard and Tarantola 1995; Coléou et al. 2005). Gonzalez et al. (2008) use a multipoint geostatistics algorithm for the simulation of facies and optimize the realization according to the seismic mismatch. Saussus and Sams (2012) propose a stochastic framework to sample facies and then rock properties in an iterative and convergent procedure. Connolly and Hughes (2016) propose a pseudo-well method based on one-dimensional Markov chains for the facies simulations. Aleardi et al. (2018) propose a method for seismic AVA inversion using Markov 2 chain Monte Carlo. Larsen et al. (2006) and Fjeldstad and Omre (2017) sample the facies from a hidden Markov model.

In this work we propose a new method to combine generative adversarial networks (GAN) with stochastic sequential simulation as model perturbation technique for facies and acoustic impedance, respectively. Both domains are simultaneously updated from a single data mismatch function.

GANs are deep generative models based on the approximation of probabilistic computations. They are composed by a specific neural network architecture with two multilayer perceptron models. One of the models corresponds to a differentiable function designated as generator (G) with inputs, from a prior input Z-vector, and hyper-parameters. The second model is a discriminator (D) differentiable function and its input corresponds to real or generated data coming from a training dataset or from G, respectively. The output of D is a binary scalar that labels the input as real or fake. Both networks are simultaneously trained in a minimax game between each other.

The joint use of GAN and geostatistical seismic inversion can be summarized in the following sequence of steps:

1. A generative adversarial network is trained to generate facies model using an ensemble of training images representing the expected geological setting.
2. After training, a latent Z-vector is used to generate facies realizations. Each model is used as conditioning data for the geostatistical co-simulation of the continuous property (e.g. P-impedance).
3. Synthetic seismic is computed from each elastic model resulting from the geostatistical simulation and compared against the observed seismic data in a trace-by-trace approach.
4. The resulting data misfit is simultaneously used to update the latent Z-vector and generate a new set of facies and used as secondary variable when updating the continuous property.

The proposed method is illustrated in non-stationary and challenging synthetic datasets with different parameterization. The different examples are used to evaluate the goodness of the method and its robustness to noise and uncertainties.

**Keywords:** Seismic inversion; Generative adversarial network; Sequential simulation; Neural network

### Geostatistical Seismic AVA Inversion with Self-updating Rock Physics

Roberto Miele*, Leonardo Azevedo, Amilcar Soares, Luiz Eduardo Varella and Bernardo Viola Barreto
*roberto.miele@tecnico.ulisboa.pt

Three-dimensional rock property models of porosity, volume of shale and fluid saturation are fundamental for reliable reservoir characterization and field development. The spatial distribution of these properties may be predicted in a two-step approach from inverted elastic models or in a single loop directly from seismic data.

The geostatistical rock physics AVA inversion (Azevedo et al. 2018) is an iterative method that allows to directly predict the spatial distribution of such properties directly from seismic reflection data by the integration of well-log data together with a calibrated rock physics model. The latter is a set of mathematical equations, which links the petrophysical and the elastic domains. There are many rock physics models available in literature to describe the relationship between such properties. These models can be as simple as empirical relations or more complex models, such as the stiff-sand and the soft-sand models (Mavko et al. 2005).

Thus, the definition and the calibration of the right rock physics model represents a crucial step for this inversion algorithm and is carried out by manually selecting the appropriate empirical parameters and models to fit the well log data. Once this step is completed, the iterative inversion starts by simulating 3D volumes of water saturation, porosity and volume of shale, through stochastic sequential simulation (Soares 2001; Horta and Soares 2010). Subsequently, a facies distribution volume is obtained from these simulations using a Bayesian classification approach with a facies classification at well location as training data. By applying the calibrated rock physics model to each facies of the volume, three models for P- and S-wave velocities (Vp, Vs) and Density are derived. Finally, angle dependent synthetic reflection seismic data can be calculated and compared against the real seismic volumes. The similarity between synthetic and real seismic traces drives the convergence of the iterative procedure both at the local and global scales. In this frame-work, the calibration of the rock physics model represents a deterministic step. This is a downside of the method that limits the range of computed Vp, Vs and Density values and therefore might impact negatively the convergence of the method and the exploration of the model parameter space.

To overcome this problem, we propose to integrate Statistical Rock Physics (Avseth et al. 2005), into Geostatistical Rock Physics AVA inversion based on the concept of self-updating joint distributions. In the first iteration a set of possible elastic responses are computed from statistical rock physics following: (i) definition of facies from well log data; (ii) rock physics modelling and Monte Carlo simulation of elastic properties (Vp, Vs and density); and, (iv) pdf estimation.

In the proposed iterative geostatistical seismic inversion, we start by simulating a set of rock properties (e.g. triplets of porosity, water saturation and volume of mineral) with geostatistical sequential simulation. Each triplet is used to compute its elastic response based on the multivariate distribution resulting from the statistical rock physics. The mismatch between the synthetic seismic, computed from the elastic models, and the real seismic is used to update the multivariate distribution initially estimated with statistical rock physics. In this way the multivariate distribution is update based on the data misfit. We show the application of the method in synthetic and real examples.

**Keywords:** Seismic inversion; AVA; Sequential simulation; Self-updating

**Ensemble Smoother with Model and Data Reduction Using Machine Learning**
Leandro Passos de Figueiredo*, Rodrigo Exterkoetter, Alexandre Anoze Emerick, Fernando Luis Bordignon, Dario Grana, Bruno Barbosa Rodrigues and Mauro Roisenberg
*leandro@ltrace.com.br

Generally, in inverse modeling in geoscience, we aim to predict the values of a group of model variables from a set of observed data, based on physical relations between model parameters and data. Specifically, in seismic inversion, the goal is to predict rock and fluid properties in the subsurface from seismic and well-log data. The relation between elastic and petrophysical properties and the distributions of these properties vary in different facies. For this reason, seismic facies inversion can be formulated as a mixed discrete-continuous problem in which the continuous rock and fluid properties depend on an underlying unobserved discrete variable representing the facies.

Ensemble-based methods have been successfully applied for data-assimilation in geosciences. In particular, the Ensemble Smoother with Multiple Data Assimilation (ES-MDA) performs smaller corrections for each ensemble update, avoiding large Gauss-Newton corrections, and can be applied to non-linear applications. However, ES-MDA is generally limited to problems where the model variables are continuous and cannot be applied, in the original form, to mixed discrete-continuous problems. Deep learning has been successfully applied in several fields for classification and pattern recognition problems. In particular, the Generative Adversarial Network (GAN) has been proposed for unsupervised training of generative models for complex distributions. Recently, the integration of the Variational AutoEncoder (VAE) in the ES-MDA approach was proposed for the estimation of facies in reservoir models based on hydrocarbon production data.

We propose to combine ensemble-based methods and deep learning algorithms for facies inversion from seismic data. In the ES-MDS process, the GAN is used to predict the discrete property, i.e. the facies classification. In particular, we use a set of Markov chain simulations to train a GAN for data re-parametrization. This approach allows generating facies realizations from a set of continuous Gaussian-distributed properties. The proposed methodology was applied to a synthetic case generated from a pre-salt reservoir model to validate the method. The application shows that the proposed methodology provides accurate results for seismic facies inversion, with data limited signal-to-noise ratio.

**Keywords:** Generative adversarial network; Seismic inversion; Deep learning

# Mining

### Application of Multiple-Point Statistics for Stratigraphic Modelling of Coal Layers

Sultan Abulkhair* and Nasser Madani
*sultan.abulkhair@alumni.nu.edu.kz

Coal deposits frequently represent complex geology comprised of seam layers with long connectivity, geological modelling of which plays a crucial role in resource estimation and further mine planning processes. Conventionally, seam layers can be modelled by wireframing technique, which may be biased as it depends on the opinion of experts. However, the accuracy of this technique depends on a dense sampling pattern of borehole data. Therefore, in a geostatistical modelling context, one of the challenging issues in the probabilistic description of coal deposits may involve selecting an optimal algorithm that can characterize the connectivity feature of the seam layers. Regarding this, variogram-based geostatistical approaches may be an alternative, yet they are highly dependent on the number of exploration boreholes to infer a reliable variogram model for characterizing the spatial continuity of underlying variables. On the contrary, Multiple-point statistics (MPS) methods have already proven their effectiveness in modelling curvilinear geological structures such as well-connected channels in petroleum reservoir settings. The main difficulty in this approach may be related to deriving a trustworthy gridded training image (TI) that should be significantly larger than the target simulated grid and a limited amount of boreholes. This research constructs a training image for three main seams based on the dense sampling pattern of borehole data in a coal deposit located in Kazakhstan. This 3D image is then applied to inform the stratigraphic layers in a part of the deposit where only three distant boreholes are available. To do so, the Direct Sampling algorithm (DeeSse) has been used, thanks to its flexibility and less requirement to high computation resources. Realizations from MPS simulation are validated by comparing to the TI in terms of reproduction of proportions, variograms and connectivity. In addition, the results of MPS are compared with realizations obtained from Plurigaussian simulation, a well-known variogram-based approach for geodomain modelling. The statistical analysis and visual inspections corroborated that MPS outperforms Plurigaussian simulation in terms of geometry reproduction of connectivity among seam layers and can be used for further resource modelling of coal in this particular sedimentary deposit.

**Keywords:** Multiple-point statistics; Training image; Coal seam; DeeSse algorithm; Coal deposit

### Geometallurgical Modeling and Deep Q-Learning to Optimize Mining Decisions[4]

Sebastian Avalos* and Julian M. Ortiz
*sebastian.avalos@queensu.ca

Mining decisions are driven by every step in the mine value chain. Integrating predictive models, at different stages of a mining complex, is necessary in order to capture the maximum mine profit, accounting for the intrinsic variability in the ore attributes and process-es performances.

We analyze how models at different stages of the mining value chain can be integrated to drive decision-making and optimize the profitability of the mine project in a closed geometallurgical framework. We compare the resulting values with the conventional case where decisions are mostly made by simplified regression models.

---

[4] A modified and extended version of this work has been submitted to the Geostats 2021 Special Issue in the journal Mathematical Geosciences.

The base case is built by using conventional estimation methods and simplified regressions models, for different ore processes. Based on a univariate estimated block model, a mine plan and schedule are used to feed the processing plant. Performance at each processing stage is modeled with conventional metallurgical techniques, based on regression from geometallurgical test work. The economic performance of the project is assessed for this case to provide a reference value.

Then we study the case where uncertainty in resources is characterized with geo-statistical methods. We consider the multivariate feature of the dataset. We compare the reproduction on grade distributions of two different frameworks for conditional geostatistical simulation. Accounting for grades variability, a mine plan is obtained using deep learning architectures trained via deep Q-learning. More sophisticated predictive models of the mine extraction and ore processing are developed to assess the production performance. The resulting economic performances are compared with the reference case, and the main drivers of value are identified and discussed.

**Keywords:** Schedule optimization; Production planning; Geometallurgy; Machine learning

## Selective Mining Unit (SMU) Study by Simulation of a Copper Deposit, Chile

Antonio Cortés Pizarro*
*acortes.mba@gmail.com

At the feasibility stages of a mining project various parameters are relevant in relation to the value of the project and must be confirmed. The selective mining unit (SMU) is a relevant parameter related to operational constraints and its definition should be supported by sensitivity analysis regarding its impact on the project value. This study measures the impact of different SMUs in terms of economic value and projected operational dilution and ore loss.

Firstly, the study involves the simulation of geology and grades and an economic assessment is performed for the different SMU's. From this first stage a given SMU size is selected, and a dilution study is performed next which involves obtaining a blast hole ("fictitious") dataset from simulations, estimating grade control models from blast holes, and applying diglines on the grade control models to quantify the projected amount of dilution for the selected SMU.

The deposit used for this study is a porphyry copper located in the Atacama Desert in the North of Chile, and coordinates have been modified to preserve confidentiality. The paper presents the procedure and results obtained.

**Keywords:** Selective mining unit; Dilution; Ore loss; Conditional simulation; Estimation; Grade control

## Chemical Anisotropy Concepts and Stockpile Management Applied for Variability Reduction in Serra do Sapo Iron Ore Deposit—Brazil

Fernando Rosa Guimaraes*, Geraldo Sarquis Dias and Cláudia Mara Sperandio Neves

*fernando.guimaraes@angloamerican.com

A performance gain on the beneficiation plant was observed in moments when iron ore was fed with low variability of iron grade and contaminants. This gain is related mainly to mass and metallurgical recoveries and the pellet feed quality. According to the variography analysis, the alumina and iron grades are more continuous along the strike direction than along the dip direction. This means that the Fe variance along the NS direction is at least 20% lower in comparison with EW direction—in case a specific geological domain at Serra do Sapo Deposit. Therefore, if the mining sequence was planned elongated on NS direction, the variability of alumina and iron grades can be reduced during the mining and plant feeding.

Some researches were carried out based on this natural phenomenon, in order to demonstrate the variability reduction if mining polygons are suitably oriented. It was developed an exercise applied inside an area with drillholes spaced by $12.5 \times 12.5$ m, comparing the standard deviation calculated from samples inside a block elongated on NS direction, against samples selected inside a block elongated on EW direction. Significant differences in variance of those groups of samples was confirmed. In addition, the behaviour of the variograms along the thickness demonstrated also high variances on very short distances, providing a positive effect on the variability decrease in case of smaller benches mining. Another exercise revealed clear NS elongated trends of iron grades with narrow ranges (from 45 to 50% Fe), while on the EW direction longer intervals were demonstrated (from 32 to 50% Fe).

Information from dispatch systems regarding daily and hourly variations in iron and alumina grades were analyzed for a full month and considering the mining progress direction for 24 h. It was confirmed that during the days when the mining progress was elongated on NS and NNW-ESE directions, the variance of grades on plant feeding was significantly smaller than on days when the mining direction was developed on EW direction or on a spread way.

Some contributions of material coming from stocks have also high influence on grade variability during the feeding, once some homogenization process occurs there. During around eleven months in Minas-Rio System, proportions of materials from stocks ranging from 25 to 62% were used for plant feeding and good correlation with mass recovery was obtained.

Therefore, the combination of mining elongated on direction of higher continuity of grades, together to good management of stockpiles may be a good point to explore in order to reduce the grade variability during the mining and consequently improve the mass and metallurgical recoveries and pellet feed quality.

**Keywords:** Ore variability; Variograms; Directional anisotropy

## Dynamic and Interactive Dashboards for Mineral Resources and Ore Reserves Management and Controls Used on Iron Ore and Nicked Deposits—Brazil

Fernando Rosa Guimaraes*, A. H. Caires Jr., P. H. Faria, B. S. Conceicao, Geraldo Sarquis Dias, Cláudia Mara Sperandio Neves, and T. M. Faria
*fernando.guimaraes@angloamerican.com

Mineral Resources and Ore Reserves documentation and reporting processes involve a huge number of information and professionals and requires to be carefully managed and controlled. Two full dashboards were designed and developed to assist on managing all information in single and dynamic workflows. Each necessary step to develop the resources and reserves estimates and their documentation can be dynamically accessed by direct hyperlinks that must be stored in a cloud provider or in a specific area on the global server.

All data regarding to mineral resources like drillholes info, assay results, QAQC reports, photos of drill cores, information of recovery, topography, mineral rights, land owners and all information about geostatistical analyses can be promptly accessed by the links. Internal and external controls and documentation are individually separated according to each stage. Some control gates were also established, and they are relevant to evaluate the data quality and feasibility of each step. Gates associated to database and geological model handovers, validation of geological interpretation, grade and density estimates (swath plots), dataset confidence level, final figures review among other were defined for step validation.

Regarding ore reserves, the same system enables the controlled (categorized users) and hastily access to the modifying factors including economics (costs, price forecast, exchange rate, discount rate, revenue factor and DCF), technical (geotechnical, dilution, dilution/ore loss, metallurgical process and routes, production plans), environmental and social topics, risks and opportunities. Data from the Terms of Reference to the final report including all traditional steps (cut-off policy, optimization, design, scheduling) can be queried in the dashboard.

Information of internal and external audits and their associated action plans, CP reports, standard operational/technical procedures, international mining codes and general guide lines, risk assessment, value chain reconciliation data, beyond CP abridged CV, signed off appointment letters, procedures of database back up among others can also be directed accessed by the links.

**Keywords:** Data management; Data integration; Ore resource; Ore reserves; Geostatistics

## Multivariate Analyses of Chemical, Geometallurgical, Mineralogical and Hyperspectral Information to Assist Optimization of Iron Ore Beneficiation Plant Performance—Brazil

Fernando Rosa Guimaraes*, G. S. Dias, R. G. Ferreira, F. P. Morais, A. D. da Rocha, E. F. Castro, B. B. O. Duarte, H. D. G. Turrer, C. M. S. Neves, C. R. S. Filho, D. F. Ducart and R. Scafutto
*fernando.guimaraes@angloamerican.com

Multivariate analysis combining geological and mineral processing information were carried out at the Serra do Sapo Iron Deposit—near Conceicao do Mato Dentro town (MG), aiming the optimization of some process inside the beneficiation plant and increase the mass recovery. An integrated database with quantitative information about minerals, metallurgy and elemental grades was developed, grouping chemical, geometallurgical, mineralogical and hyperspectral samples, and analyses from multiple campaigns of reverse circulation and diamond drilling. A big matrix with correlation indexes of 74 variables was calculated in order to get a better understanding about the characteristics of the iron formation lithologies related to the plant behaviour.

The spatial distribution of the all variables, estimated in block models by several interpolation methods, and the highlighted pairs of key variables with high correlations, enabled to create dashboards with multiple features of the ore that will be mined in monthly and weekly scales, providing, in advance, relevant information that will be useful for the beneficiation plant team manage and calibrate the process controls related to the specific material that will be fed. In addition, this multiple information can be useful for blending strategies and storage impact in iron ore quality, focusing on plant performance improvement. The spatial distribution of some key variables is also profitable for guiding the mine planning team to establish strategies for mining sequence focusing in manage specific periods for Direct Reduction or Blast Furnace production, based on some really key variables and depending of the market strategy.

Some highlighted points were observed and had important influences on plant performance. In areas with predominance of lamellar hematite provides good mass recovery on flotation; good correlation between grain size, iron recovery on desliming and work index; high correlation between selectivity index and iron recovery on desliming, high negative correlation between kaolinite, calculated by hyperspectral methods, with mass recovery on flotation, etc. Strategies of stockpiles framing and online blending with material from different mining faces starts to be developed, considering the spatial distribution of the geometallurgical, mineralogical and hyperspectral variables in addition with chemical ones, in a synchronized way according to periodic demands.

**Keywords:** Multivariate data analysis; Correlation matrix; Stockpiles; Blending

## Effects of High-Order Simulations on the Simultaneous Stochastic Optimization of Industrial Mining Complexes

Joao Pedro de Carvalho* and Roussos Dimitrakopoulos
*joao.decarvalho@mail.mcgill.ca

An industrial mining complex or mineral value chain is an integrated business that includes mines, stockpiles, waste/tailings dumps, processing streams and related facilities, leading to the generation of products delivered to customers and/or the spot market. Supply variability and uncertainty of the materials extracted from the related mines are quantified with geostatistical simulations providing the inputs to the simultaneous stochastic optimization of mining complexes. The effects of utilizing traditional Gaussian approaches with their maximum entropy properties as opposed to distribution assumption-free and spatially more informed multiple-point or high-order approaches on the results of the related optimization is a point of practical interest. In this paper, the effects of using sequential Gaussian simulation versus the sequential high-order direct block simulation in the simultaneous optimization of a gold mining complex are explored. The later complex is composed of a gold mine, leach-pad, stockpile, waste dump and processing plant. Results show that the high-order simulation approach, generating realizations with more realistic connectivity of high-grades, results in a more informed optimization process and better life-of-mine production schedule, with a net present value increase of 5–16%, when compared to the life-of-mine generated based on the sequential Gaussian simulation. Notably, the extraction sequence is driven towards areas where the high grades are spatially better connected, leading to both higher head grades feeding the related processing plant and a smarter extraction sequence extracting less waste.

**Keywords:** Sequential Gaussian simulation; Direct block simulation; Connectivity; Production optimization

## Mining Dilution and Stockpiling Study by Simulation of a Gold Deposit, Canada

Georges Verly* and Henry Kim
*georges.verly@woodplc.com

Cote Gold is a large Archean gold porphyry style deposit amenable to open pit mining at the development stage. The value of the project depends on stockpile sequencing during production allowing to process higher grade ore first. The project owners had the following questions: (1) what the ability of the mine plan is to deliver predicted tonnes and grades to the mill and stockpiles to meet the production plan; (2) what the ability of the estimated resource block model is to predict the various recovered tonnes and grades; and (3) what the impact of different blast hole sample preparation protocols on the ore/waste misclassification during mining is. These three questions were addressed with two conditional simulation exercises during prefeasibility and feasibility stages.

The simulation and post-simulation procedures used consisted of generating several realizations of the gold grade on a tight grid using Sequential Gaussian Simulation, selecting composite values from the realizations and adding a relative noise to get simulated blast hole datasets, estimating grade control models using the blast hole datasets, applying dig-lines on the grade control models to segregate the various waste, stockpile and mill ore categories followed by visual and statistical assessment aimed at answering the owners' questions.

This paper present details on the procedure and results obtained, including validation and rational for the multiple decisions taken during the study.

**Keywords:** Conditional simulation; Sequential Gaussian simulation; Grade control; Dilution

### Resource Assessment of Copper Mine Tailings

Fabian Soto*, Felipe Navarro, Brian Townley, Manuel Caraballo, Patricio Martinez and Rene Martinez
*fsoto@alges.cl

Several tailings were built at a time when recovery was focused only on one element of interest, when geological knowledge of the deposit was not aware of oxides-sulfides; and, even sometimes, where cutting grade was higher than nowadays cutting grade. These tailings were built from different mine residuals and from years they are in abandoned sites. During the last few years, they have gained increasing interest due to technologies for drilling samples, understanding and evaluating chemical stability and to evaluate the commercial interest of extraction of ore minerals.

A resource assessment approach to address and quantify the available resources was applied here. Different sampling techniques used to extract drillholes are explained, and it is illustrated by the estimation of three cases of tailing containing at least five elements of interest among: copper, magnesium, phosphorus, iron, cobalt, gold, silver and rare-earth elements.

**Keywords:** Uncertainty; Resource evaluation; Tailings

### A Review of Applications of Geostatistical Simulation Models in Remote Sensing

Joao Neves*, Diogo Cauppers and Amilcar Soares
*vermelho.neves@tecnico.ulisboa.pt

Mining resources are normally characterized in a block model in which the mining area is discretized. The underlying uncertainty associated with the grade values of mining blocks (block uncertainty) can be assessed through stochastic simulation methods, which allow the characterization of local pdfs of grades. After the characterization of mean grades and uncertainty per block/stope, the main role of mine planning consists on characterizing the time scheduling of production from reserves in terms of a mining sequence.

The challenge lies in transferring block uncertainty into a temporal flow of mean grades, and the consequent grades' uncertainty at time $t$. The most straightforward approach consists of calculating an optimal mining sequence (by minimizing an objective function) and to apply this to each simulated block model, in order the uncertainty of each period can be assessed.

However, this approach needs to compute and retain the $N$ simulated models. This can be a cumbersome task, particularly for very large block models. Also, any time we have new sample data, taken at the stopes during production, the entire ensemble of simulated models needs to be re-evaluated.

These are the main reasons why most mines do not use stochastic simulation for uncertainty assessment in their short and medium-term mining planning routines.

This work proposes to tackle these two issues with simple implementation methods: integration of uncertainty in dynamic time scheduling, and fast updating of stochastic resources models. Both methods rely on firstly converting the ensemble of stochastic simulated models of grades into a few models describing mean grades and quantiles.

- In the proposed methodology the short mining schedule is computed using not only the mean grades, but also using the uncertainty as a factor in the optimization process. This is accomplished by aggregating the uncertainty of several blocks/stopes, to be mined in a given period, via a non-parametric method of interpolation of pdfs. Once the uncertainty of grades, of a set of stopes, of a given period of the time scheduling is known, it can be used as an optimization parameter either in the context of internal blending strategy or in a selective mining.
- As for the fast updating of the models, with new sample data, a new method of fast stochastic simulation update is proposed, which allows for local update of the quantiles used by the mine planning for the optimal temporal scheduling characterization.

Both methods are being implemented in a real case study and preliminary results are presented.

**Keywords:** Mine planning; Production scheduling; Conditional simulation

### Mineral Resource Classification Using Machine Learning

Ilkay S. Cevik*, Oy Leuangthong, Antoine Caté, David Machuca-Mory, Julián M. Ortiz
*icevik@srk.com

Mineral resource classification is a subjective task, performed near the end of the mineral resource modeling workflow. Three categories of mineral resources (Measured, Indicated, and Inferred) are used to distinguish between different levels of confidence reflecting the quality of the data, geological interpretation, resource estimation, and the reasonable prospects for eventual economic extraction. There are no prescribed approaches to assign classification; the criteria and way it is implemented is left to the discretion of a Qualified or Competent Person, as this term is defined by regulatory bodies.

This paper presents a methodology that consistently integrates multiple sources of information that are commonly associated with one of the four factors considered for classification: data quality, geological confidence, resource estimation metrics, and reasonable prospects for eventual economic extraction. The aim is to assimilate both quantitative and qualitative data, often available in the form of a 3D block model, in a consistent and repeatable fashion. Consistency and repeatability are both important objectives since mineral resource models are often updated at least annually, and barring no significant changes in geologic interpretation, there should be some consistency in how resource confidence is defined from one year to the next. In addition, the approach should save the modeler professional time, especially in subsequent model updates.

The general methodology involves two primary steps. The first step is to cluster blocks with similar parameters or inputs to obtain an initial classification category. The second step is focused on smoothing of the categories to ensure continuity of classified blocks, which is particularly important for higher confidence or Measured blocks.

The first step of defining an initial classification category depends on the type of data available. Two workflows are presented, depending on whether the factors considered are solely quantitative or a combination of quantitative and qualitative metrics.

If all inputs are quantitative, an unsupervised clustering approach can be considered; however, given the large size of block models today, this direct approach is both time and memory intensive. A practical solution is to subsample the block model and run the unsupervised random forest algorithm to obtain a distance matrix, which is then used to as training data to cluster the block model via a supervised random forest. This subsampling, and combination of unsupervised and supervised RF is repeated many times to calculate the probability of belonging to each class at each location. These class probabilities are used to assign the initial classification category.

When qualitative information is available, a numeric score is given to both quantitative and qualitative criteria where a higher score implies higher confidence level in the geology interpretation, grade estimation, data quality and/or the reasonable prospects for the eventual economic extraction. Weights are assigned to each criterion at the discretion of the expert, and a weighted average score is calculated. The final scores are clustered to obtain an initial classification. The second step in the process involves smoothing of the boundaries between resource categories to ensure continuity of classification categories and avoid isolated categorized blocks. For this purpose, support vector classification (SVC) with a radial basis function kernel is proposed. The final block classification is obtained by tuning the SVC hyperparameters.

Two applications for two separate gold deposits are presented to illustrate the methodology. The first example uses only quantitative data, while the second example considers both quantitative and qualitative inputs. In both cases, results are comparable to the classification done by the project qualified person using conventional methods.

**Keywords:** Resource classification; Random forests; Machine learning; Supervised learning

## Risk Assessment of Mining Dig Lines in a Multivariate Mineral Deposit with Sum and Fraction Constraints

Jonas Kloeckner*, Joao Lucas de Oliveira Alves, Marcel Antonio Arcari Bassani and Joao Felipe Coimbra Leite Costa
*jonas.kloeckner@ufrgs.br

Mineral deposits often consist of a complex arrangement of multiple related variables and in some cases have non-linear relationships. Further intricacies may include fraction and sum constraints. Fraction constraint means that one variable may not exceed another, and sum constraint means that the sum of some variables may not exceed a constant. Geostatistical simulation has been widely used for risk analysis in mining planning. In the case of multiple correlated variables that have sum and fraction constraints, geostatistical simulation is a challenge, as the constraints and relationships among the variables must be reproduced in the final models. This study applies multivariate geostatistical simulation with constraints to assess the risk associated to the grades predicted in mining polygons in a real bauxite deposit. The workflow consists of the following steps: (i) transforming the original variables into log-ratios, (ii) transforming the ratios into independent Gaussian variables using the Projection Pursuit Multivariate Transform (PPMT), (iii) simulating the Gaussian variables independently through sequential Gaussian simulation and (iv) back-transforming to the original variables. The realizations were checked and reproduced the variograms, histograms, multivariate relationships and constraints. The realizations were performed at point-support and upscaled to the volumes of the dig lines used in the mine. Finally, the simulated realizations of the dig lines were used to measure the dig lines with higher risk of not achieving the required product specifications.

**Keywords:** Multivariate simulation; Fraction constraint; Sum constraint; Projection pursuit multivariate transform; Risk assessment

## Combining Indicator Kriging and Geostatistical Simulations to Classify Mineral Resources: A Major Bauxite Deposit Case Study

Octavio Rosa de Almeida Guimaraes*, Flavio Henrique Tavares da Silva and Joao Felipe Coimbra Leite Costa
*flavio.silva2@alcoa.com

Mineral resources are usually classified as measured, indicated and inferred according to their associated degree of confidence. There are several international codes for reporting resources and reserves recognized by the committee for mineral reserves international reporting standards (CRIRSCO) which define and suggest guidelines to classify the ore. Even though the final criteria adopted is a personal decision defined by the so-called competent (qualified) person who must have proven experience related to style of the mineralization. The mineral resource statement includes the quantity of ore usually expressed in mass and its quality (commonly grades). Geostatistical methods provide the means to access the level of uncertainty associated with these quantities and qualities declared. Various geostatistical frameworks had been adopted to access the uncertainty regarding these values. This specific paper presents a possible solution combining indicator kriging (IK) and geostatistical simulations. The first is used to measure the uncertainties related to the definition of the ore zone in terms of volume. The second used to quantify the uncertainty regarding the grades and metal contents. The methodology proved efficient and is used on a major bauxite mine north of Brazil. The case study illustrates the proposal which is capable to define parts of the deposit with decreasing levels on confidence related to each class of mineral resources.

**Keywords:** Resource classification; Indicator kriging; Conditional simulation

## Generative Adversarial Networks for Improving Geostatistical Models in Complex Orebodies

Helga Jordao* and Amilcar Soares
*helga_jordao@hotmail.com

Mineral resources evaluation is highly dependent on the spatial characterization of the different ore type domains. In complex geological environments there is still a need for an expert geological control to perform the characterization of these models since using quantitative methods remains a challenge. The need for human control is time-consuming which is an impediment for geological resources to be frequent updated.

In this paper a deep learning method, Generative Adversarial Network, is proposed to mimic the geological interpretation of one or more geologists. Based on a set of bore hole samples and the consequent envelope of the different ore types resulting from the expert geological interpretation, the model assumes the images of the interpreted ore types as reference images and the bore hole samples as a conditioning starting image.

The proposed method is applied to a real sulphide deposit located in the south of Portugal. The results have shown the deep learning model was capable of learning the underlying patterns of a specific complex geological environment and transfer the learned knowledge to generate new geological models. We consider that this approach presents promising results for its application on real-time geological model management.

**Keywords:** Generative adversarial network; Deep neural network; Orebody modeling

## Using Multiple-Point Statistics Through Estimation to Decide Where to Drill

Oli D. Johannsson*, Mats Lundh Gulbrandsen and Thomas Mejer Hansen
*oli.johannsson@geo.au.dk

One of the main goals of applying geostatistical modeling is to allow decision makers to take decisions based on incomplete knowledge of the subsurface. The geostatistical model itself conveys the expected uncertainty and variability of the subsurface. In addition, other types of information are available, that can be included in the geostatistical model, such information from well logs, and geophysical measurements.

Given a data set based on a geostatistical model and a number of well log measurements, we consider the specific problem of: Where should the next borehole be drilled, in order to maximize the information in the model? One way to tackle this problem is to perform conditional geostatistical simulation, in order to generate a large set of subsurface models. Then the variability in each cell can be computed, from which one can get a measure of where information is lacking and hence where to drill next. As 3D geostatistical simulation, and especially multiple-point statistical (MPS) based simulation, can be computationally challenging, such an approach can be cumbersome.

We propose an alternative approach, based on two recent developments of MPS simulation: Well log data are in general uncertain, and typically only co-located uncertain data are used in MPS based simulation. Here we make use of both co- and non-co-located uncertain data. Instead of simulating 100s of 3D realizations from which a measure of variability is computed, we make use of MPS based estimation that directly estimates such 1D statistics, without the need for running multiple costly MPS simulations.

We suggest a methodology where the point-wise entropy in a 3D model is computed using the recently developed MPS estimation method, conditional to the uncertain well log data. The entropy is computed directly from a 1D conditional distribution obtained from the estimation and provides a local estimate of information. The information at any X-Y location is obtained by vertically integrating the 1D entropy at each X-Y location. From the constructed 2D map of cumulative information content the location with least information content is chosen as the optimal location for the next borehole location.

We demonstrate the methodology and simulate an optimal strategy to position a number of new bore holes.

We also compare to the traditional, and computationally much more demanding approach, of using MPS simulation as opposed to MPS estimation.

**Keywords:** Estimation; Entropy; Multiple-point statistics; Optimal drilling

# Earth Science

### A Review of Applications of Geostatistical Simulation Models in Remote Sensing

Fatemeh Zakeri* and Gregorie Mariethoz
*fatemeh.zakeri@unil.ch

Observations of our planet using reflected or emitted electromagnetic energy provide in-formation to be used in different subjects such as soil, vegetation, topography, and atmospheric science. Although the number of satellites has been increased during the last decades, these data sets are often spatially incomplete, or temporal/spatial resolutions are insufficient. As geostatistical simulation models allow data uncertainty to be propagated by generating realizations, they are practical tools to fill these gaps. In this study, we review geostatistical simulation models that have been applied to satellite images, and we categorize their applications in mapping, accuracy assessment, downscaling, sampling design, and gap filling.

Mapping and classification are essential for managing natural resources. Because of their time consumption and expense, traditional mapping methods such as field surveys cannot be used for large spatial and temporal scales. However, remotely sensed data provide a practical and economical way to study different land covers. The spatial variability of observed variables can be reproduced using geostatistical simulations by generating multiple realizations, conditional to available data. Accordingly, simulations have been used in mapping a wide variety of variables such as vegetation and soil cover maps using satellite information.

The validity of the information obtained by satellite is called accuracy assessment. For instance, the fitness of a classified map derived from satellite data to the actual class label is essential for resource management. Geostatistical simulation algorithms have been used to infer uncertainties regarding landcover patterns obtained from satellite and field data by generating conditional realizations.

The subpixel information (i.e., downscaling) is needed in geoscience applications, as the resolution of satellite imagery is often insufficient. Geostatistical approaches are important tools for downscaling as they can quantify uncertainty in sub-pixel mapping. Downscaling using geostatistical algorithms can be grouped into two main approaches. One approach is to use a pair of high-resolution and low-resolution images that provide a correspondence between resolutions to generate realizations with improved resolution. Another one is to use a geostatistical simulation model to determine the probability of class labels for each pixel based on a coarse resolution classified map and a sparse set of class labels at some informed fine pixels.

In situ data is one of the information that should be used to interpret the information from satellite observations. These data sets can be used as training data for classification. It is essential to ensure the success of in situ data acquisition campaigns. One application geostatistical is to design cost-efficient sampling schemes, as these approaches can consider spatial variability and capture uncertain features across realizations.

Last but not least, geostatistical simulation models have also been used to fill the gaps that often occur in satellite data due to obstructions, such as clouds or clouds shadows. As interpolations techniques mainly can create unrealistic spatial patterns and generally do not provide uncertainty quantification, geostatistical simulation models can be used in gap-filling by preserving spatial patterns in the realizations.

Our review shows a strong potential of geostatistical simulation approaches in deriving spatial information. However, there is an untapped potential in applying these models to other domains such as change detection and information fusion. Moreover, more research is required to tailor the existing geostatistical simulation algorithms to specific remote sensing applications.

**Keywords:** Remote sensing; Conditional simulation; Accuracy assessment; Downscaling

## Probabilistic Volcanic Hazard Estimation Up to a Timeframe of 1 Ma with Assimilation of Tectonics and Geophysics

Olivier Jaquet*, Christian Lantuéjoul and Junichi Goto
*olivier.jaquet@in2earth.com

Many industrial regions around the world are concerned by volcanic risk evaluation; in particular, the Japanese archipelago due to its tectonically active nature. For risk assessments related to the isolation of potential geological repository sites, the quantification of long-term volcanic hazard becomes of fundamental importance.

For potential sites near volcanically active regions, long-term volcanic hazard constitutes the dominant source of uncertainty as input for risk assessment studies. Uncertainty is mainly related to an imperfect knowledge of volcanic processes, to space-time variability of distribution and intensity of volcanic events, as well as to a limited amount of data available.

In Japan, regions that are not obviously excluded on the basis of recent and current volcanism will be considered for the siting of a geological high-level radioactive waste (HLW) repository. The probabilistic methodology for volcanic and tectonic hazard assessment, developed by the Nuclear Waste Management Organization of Japan (NUMO) with an international team of geoscientists, addresses timeframes up to one million years. As part of this methodology, several stochastic models were developed using specific geological conceptualizations based on various data sets and information related to past and current volcanic activity, related tectonics and their geophysical signature.

For the estimation of volcanic hazard for the one million years timeframe, a representative range of plausible regional evolution scenarios was assessed in relation to plate tectonic dynamics. In particular, arc volcanism is likely to migrate within the region of interest; such effects must be taken into account when making forecasts of volcanic events for such long period into the future. Within the framework of the Cox process, the potential of volcanism is described using an evolution equation, and thus becomes non stationary in the space-time domain. Stochastic simulations of patterns of future volcanic events are function of past volcanism location, migration velocity, geophysical data and time scale considered. The estimation of volcanic hazard represented in form of maps is performed by Monte Carlo simulation, for various scenarios with timeframes ranging from one hundred thousand up to a million years.

Illustrations are provided using data from the regions of Kyushu, Chugoku and Tohoku. We emphasize that these case studies are only used as methodological examples; no region in Japan is yet considered specifically as a HLW candidate site.

**Keywords:** Risk assessment; Cox process; Data integration; Spatio-temporal modeling

## Optimizing Graduate Student Descent: Effectively Formulating Sparse Spatial Problems for Machine Learning Algorithms with Lessons Learnt from Three Applications

Jeff Boisvert*, Camilla da Silva, Liam Bennett and Dale Schuurmans
*jbb@ualberta.ca

Machine learning methods are becoming increasingly popular in many scientific fields as they offer data driven approaches to solving complex nonlinear problems. The field of Earth Sciences is no exception. Until the use of machine learning in sparsely sampled spatial problems becomes a mature application of trusted algorithms, we should take Nassim Nicholas Taleb's advice *"Learn to fail with pride - and do so fast and cleanly. Maximize trial and error - by mastering the error part"*. This presentation provides suggestions to help you master the *"error part"* of applying machine learning methods to spatial problems, with sincere apologies if you are a graduate student embarking on a machine-learning-assisted journey through "graduate student descent".

The video presentation is subdivided into three largely independent 5–10 min mini-presentations that could be viewed on their own if you are only interested in one of these topics. The first video explores the impact on several fields that have largely been replaced by machine learning, while discussing issues critical to sparse spatial problems; the second video presents three novel spatial applications of machine learning techniques to provide a context for recommendations; and the final video explores practical recommendations for machine learning problem formulation, algorithm selection, parametrization and inputs/outputs to help reduce the "error part" of trial-and-error.

**Mini-Presentation 1**: Many fields of study have been significantly impacted by data driven machine learning approaches; classical approaches to automatic speech recognition, machine vision, computational natural language understanding, and machine translation have all been supplanted by machine learning algorithms. Differences and similarities between Earth Sciences and these fields are explored. The differentiating characteristic is the degree of data sparsity found in Earth Science problems. Many mining and petroleum data sets are characterized by a large number of highly correlated and potentially biased samples that represent only a small volume of the domain, often less than a trillionth of the total population.

**Mini-Presentation 2**: For the purposes of examining spatial applications, it is convenient to characterize problems based on data density. Data driven approaches perform best with dense data, indeed, many of the aforementioned scientific fields have exhaustive data sets. Dense data is available in some spatial Earth Science applications such as remote sensing, aerial mapping, and GIS. Three novel machine learning implementations are presented that span the spectrum between sparse and dense data: (1) identification and classification of tree stands for wildfire growth modeling using exhaustive UAV and satellite imagery (2) spatial prediction of probabilistic variables for grade control with dense blast hole data (3) prediction of well decline curves from sparse data in an unconventional reservoir.

**Mini-Presentation 3**: Practical recommendations for problem formulation, algorithm selection and parametrization are provided. This work provides practical recommendations derived from the authors' significant experience with machine learning algorithms as well as various case studies. These specific recommendations are generalized into insights into the future of machine learning in the Earth Sciences based on the review of how machine learning has impacted other fields.

**Keywords:** Machine learning; Earth sciences; Sparse data

### Unsupervised Modelling of Geoscientific Data via a Spatially Aware Random Forests Algorithm

Hassan Talebi*, Luk J. M. Peeters and Alex Otto
*hassan.talebi@csiro.au

Unsupervised learning helps to find previously unknown patterns in geoscientific datasets without pre-existing labels. Non-spatial learners generally look at the observations based on their relationships in the feature space, so they do not have the means to consider spatial relationships between the regionalized variables. This study introduces a novel spatial random forests technique based on higher-order spatial statistics for unsupervised modelling of spatial data. The dimension of the input training data is increased by adding information from the neighboring locations. A synthetic dataset is generated from the original data via shuffling the locations of the surrounding information followed by random sampling of the marginal distributions (destroying statistical and spatial information). The random forest is trained to discriminate the original data from the synthetic ones. The proximity between original observations is measured by the frequency of sharing the same terminal node in the spatial decision trees. The final spatially aware proximity matrix can be used for clustering the input data. The superior performance and usefulness of the proposed algorithm are illustrated via one synthetic and one real case study, where the geophysical and remotely sensed covariates in Yilgarn craton (Western Australia) are used as input data for geological mapping and process discovery analysis.

**Keywords:** Spatial random forests; High-order spatial statistics; Unsupervised learning; Domaining; Spatial data

**To MPS or Not to MPS—Turning a Geological Interpretation Model into a Probabilistic Geological Model** Rasmus Bødker Madsen*, Ingelise Møller and Thomas Mejer Hansen
*rbm@geus.dk

Combining geophysical data with information such as from boreholes and expert domain knowledge is nontrivial when generating geological models of the subsurface. One common solution is to deterministically invert the geophysical data and grid it alongside e.g. boreholes in a visualization tool. This allows a skilled geologist to rationally interpret the geophysical data as geological units, formations etc. and hence implicitly combine the geophysical data with geological knowledge and borehole information. This procedure is known as cognitive geological modeling. Several potential sources of uncertainty are present in cognitive modeling. From the initial measurement errors on the geophysical instruments, to the way the data is gridded, inverted, and processed, and finally in the combination process of the actual interpretations. The final output of cognitive geological modeling is a single model of the subsurface. Handling the described uncertainties is therefore difficult if not impossible. These types of models consequently are also sensitive to bias that accumulates through each step of the modeling procedure.

To remedy the shortcomings of this strategy we propose a novel stochastic methodology combining the efforts of probabilistic data integration and cognitive modeling. We treat geological interpretation points from the cognitive model as uncertain "soft" data. This data is then combined with analogous geology in a probabilistic model. We test three ways of combining and sampling from such a probabilistic model.

Firstly, two maximum entropy setups based on a two-point statistical approach. One based on sequential Gaussian simulation (SGS) and one based on Cholesky decomposition of a Gaussian covariance. Secondly, lower entropy (and conceivable more realistic) geological solutions are obtained from multiple-point geostatistics (MPS). We apply both ways of solving the problem at a study site near Egebjerg, Denmark, where Airborne TEM, seismic data, and borehole information are available and interpreted in cognitive modeling. Results show that both the two-point statistical and multiple-point statistical approach allows satisfactory simulations of uncertain geological interpretations and are consistent with prior geological knowledge. MPS simulations are usually more computationally expensive than Gaussian simulation. So how does one choose and adapt an appropriate scheme and avoid cracking a small geostatistical nut with a MPS sledgehammer? Our results show that the number of soft data points play a pivotal role in answering this question. MPS simulations allow connectivity in scenarios with few data points due to the low entropy of the model. Conversely, when the number of soft data increases, SGS is less prone to produce simulation artifacts due to inconsistencies with the prior and is therefore advantageous.

**Keywords:** Multiple-point statistics; Sequential gaussian simulation; Probabilistic data integration

# Domains

### High-Order Geostatistical Simulation of Geological Units: Application at the Saramacca Gold Deposit, Suriname

Daniel Morales* and Roussos Dimitrakopoulos
*daniel.morales2@mail.mcgill.ca

A sequential, data-driven, high-order categorical simulation method (HOCSIM) is used in this paper to simulate geological domains at the Saramacca gold deposit in Suriname. Unlike multi-point simulation approaches, HOCSIM does not require the use of training images and it is based on high-order spatial indicator moments that consistently relate low and high-order moments via boundary conditions. For the sequential simulation process, a recursive B-spline approximation algorithm is employed leading to the reconstruction of the high-order spatial statistics of the available data in the simulated realizations. The step-by-step application of HOCSIM at the Saramacca gold deposit demonstrates the related practical aspects. Given the six nested geological domains of the Saramacca deposit, a hierarchical implementation of the method is adopted. In the first step, two units are generated reflecting the identified mineralized and non-mineralized deposit interpretations. Then, a second pair of categories are simulated corresponding to the Pillow Basalt and Fault Zone domains within the mineralized part, simulated in the first step. Finally, three domains are simulated within each of the two domains from the previous step. Validation of the simulated realizations demonstrate the reproduction of data statistics, namely, proportions, spatial indicator variograms, and third- and fourth-order spatial indicator moments.

**Keywords:** Categorical simulation; Geologic domains; High-order moments

## 3D MPS Joint Simulation of Geology and Redox—An Applied Example from Denmark

Rasmus Bødker Madsen*, Ingelise Møller, Hyojin Kim, Anders Juhl Kallesøe, Peter Sandersen and Birgitte Hansen
*rbm@geus.dk

Nitrate contamination in subsurface aquifers is an existing issue due to intensive nitrogen fertilization and management in agriculture. As nitrate is moving through the subsurface via water, it is reduced only in the reduced zone coupled with oxidation of naturally occurring geological material such as organic rich clays and pyrite. To assess the state and vulnerability of aquifers and its potential as a water resource, spatial information of both the water pathways and the redox conditions along the pathways are therefore essential. The flow path of the groundwater is primarily governed by the geology and the fate of nitrate is determined by the redox, and reduction rates of the subsurface. Redox conditions are traditionally mapped by delineating an interface between oxidizing and reducing conditions, while geology is modeled separately either in cognitive modeling or through geostatistical simulation.

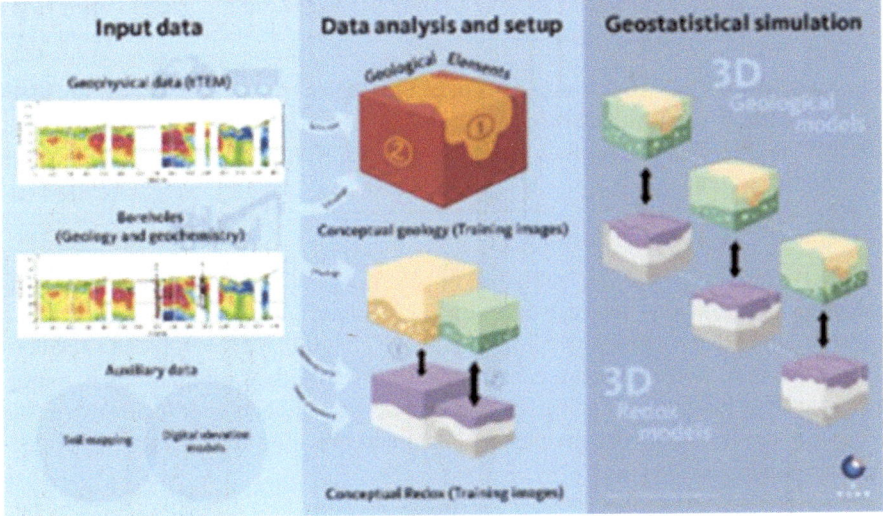

We propose a novel approach to model both redox structure and geology simultaneously in high resolution 3D (25 m × 25 m × 2 m) using multiple-point geostatistical simulation (MPS). The proposed workflow is illustrated in previous figure. Boreholes containing 1D densely sampled information about geology and sediment color, which is translated into redox conditions, are available. Geophysical data, in the form of land based towed transient electromagnetic measurements (tTEM), provides the spatial 3D architectural information. These data are combined with detailed soil maps and digital elevation models to identify main geological elements in the area. These elements are from a geological perspective considered independent. This allows a computationally attractive solution as the simulation can be subdivided into the smaller geological elements instead of the area as a whole. Our working hypothesis is that geology and redox are coupled within each geological element. Hence, a bivariate training image is constructed for each element. Boreholes and soil map data are introduced to the modeling domain in the form of hard data. Geophysical data are translated into soft probabilities of geology. The final realizations stitch together the simulations of the geological elements as shown in the previous figure.

The ensemble of realizations represents a quantification of the uncertainty for the given setup. MPS modeling allows mapping of complex connectivity in the geological domain. In contrast to traditional redox surface modeling, our approach introduces a more multi-faceted description of the subsurface. The introduction of training images makes the incorporation of expert geological and geochemical knowledge easy and intuitive. Importantly, the use of bivariate training images ensures that redox and geology realizations are consistent. This is crucial for a precise hydrological modelling of the fate of nitrogen from the root zone to groundwater and surface waters.

**Keywords:** Multiple-point statistics; Hydrological modeling; Nitrate contamination; Conditional simulation

## Resistivity-Lithology Relations: A Tool in Geological Modelling Based on Electrical and Electromagnetic Data

Ingelise Møller*, Rasmus Bødker Madsen, Anders J. Kallesøe, Peter B. E. Sandersen and Thomas M. Hansen
*ilm@geus.dk

Working with geological interpretation of electrical and electromagnetic data it is important to know how the resistivity of the involved lithologies or rocks are distributed. This relationship is affected, and hence complicated, by soil and rock conditions and even pore fluid. Information on resistivity-lithology relations is important, whenever it comes to cognitive or stochastic geological modelling or stochastic inverse modelling involving geophysical data and resulting in an ensemble of geological models.

Resistivity-lithology relations have over the years been studied in relation to local or regional scale surveys, where the locally observed resistivities obtained by electrical and electromagnetic surveys or borehole wire-line logging are compiled with local geological well logs. There are only few examples on large-scale studies. When it comes to stochastic geological modelling the simple solution is to combine the nearest geophysical sounding and borehole lithological log or look up the resistivity in a 3D resistivity grid at the position of the lithological log. If parts of the modelling area are modelled manually in a cognitive manner, i.e. for construction of a 3D training image, the resistivity-lithology relation can be constructed combining the 3D resistivity grid and the voxels in the 3D lithological grid. The resistivity-lithology relation then acts as the link between the resistivity data used as soft data and the lithological classes used in the stochastic modelling setup.

In this paper we use resistivity-lithology relations established on local, regional or country scale based on archives or databases with borehole wire-line resistivity logs and bore-hole lithological logs, where resistivity measurements can be related directly to specific and well described lithological samples. We use the Danish national geophysical database containing resistivity wireline log and the geological and hydrological database containing the lithological sample descriptions. The procedure implies a restricted use of wire-line logging data resulting in resistivity distributions for specific lithologies or geological formations. Quality controlled and documented high-quality data ensures reliable results, reflecting the actual resistivity of a specific lithology. The method is flexible when it comes to defining, which lithologies or classes of lithologies that are included in the resistivity-lithology distributions.

For computational reasons, stochastic modelling or inversion procedures typically run simplistic geological setups grouping many lithologies into a few classes. The grouping of classes is typically carried out geologically taking for instance the grain-size into account, so that clay-rich lithologies and sandy lithologies enters different classes which also coincide with low and high permeabilities, but other grouping may be used as well. However, these classes may not be optimal when it comes to the resistivity distribution of these classes. If lithologies with similar resistivity goes into different classes, the resistivity distributions will have great overlap and thereby imply similar probability in the simulations.

We explore how the grouping of the lithologies can be carried out resulting in classes with as little overlap in resistivity as possible. We both work with manual grouping and grouping applying clustering. The different grouping approaches are applied in two to three study sites with the purpose to see the impact on the ensemble of realizations obtain by multiple-point statistical simulation setup or stochastic inversion.

**Keywords:** Stochastic inverse modeling; Petrophysical data; Data integration.

# Author Index

© The Editor(s) (if applicable) and The Author(s), 2023
S. A. Avalos Sotomayor et al. (eds.), *Geostatistics Toronto 2021*, Springer Proceedings
in Earth and Environmental Sciences, https://doi.org/10.1007/978-3-031-19845-8